Design in Modular Construction

Design in Modular Construction

Mark Lawson ■ Ray Ogden ■ Chris Goodier

CRC Press
Taylor & Francis Group
Boca Raton London New York

CRC Press is an imprint of the
Taylor & Francis Group, an **informa** business

A SPON PRESS BOOK

CRC Press
Taylor & Francis Group
6000 Broken Sound Parkway NW, Suite 300
Boca Raton, FL 33487-2742

First issued in paperback 2019

© 2014 by Taylor & Francis Group, LLC
CRC Press is an imprint of Taylor & Francis Group, an Informa business

No claim to original U.S. Government works

ISBN-13: 978-0-415-55450-3 (hbk)
ISBN-13: 978-0-367-86535-1 (pbk)

Visit the Taylor & Francis Web site at
http://www.taylorandfrancis.com

and the CRC Press Web site at
http://www.crcpress.com

Contents

Foreword

The dream of industrially produced homes and buildings has inspired architects and engineers for a very long time. It was reflected in Buckminster Fuller's Dymaxion house, where similarities with the iconic Airstream caravan are obvious. The post–World War II era saw housing shortages and the emergence of novel forms of construction to meet this new challenge. Alongside conventionally produced housing built in an era of austerity, designers experimented with housing fabricated in factories or assembled from industrial components (for example, the Eames House), which relied on a range of novel structural and envelope technologies. Many approaches were highly successful.

In many ways, however, to think of modular construction as a structural or architectural approach is to miss the point. It is a means of delivery that can favour certain building design concepts, but most importantly, it is a procurement process. Modular off-site construction, whether based on volumes, components, or any hybrid system, allows a greater proportion of the construction to be moved from site to controlled environments. In these environments the timescales and economics are different, and the quality can be, too; producing buildings off-site has many advantages.

Modular construction is not so much a statement of style as it is a way of thinking about construction. It challenges many of the inefficiencies of conventional approaches. Certainly it seeks to bring the construction of buildings onto a more sophisticated footing. It feels appropriate to a world where mass production is the norm and quality and efficiency are keys to competitiveness. It is a technology that commands respect but is also one that the construction industry is only just beginning to come to terms with, at least on a large scale. The projects and technologies reviewed in this book represent the pinnacles of practice to date; they are better than their predecessors by a phenomenal margin. Moreover, there is every reason to suppose that the generations of technology that will evolve from the current position will be even more prolific and far reaching. There seems to be little doubt that this is a technology coming of age.

Christopher Nash
Nash Architecture and former Managing Partner,
Grimshaw Architects LLP

Preface

Modular construction provides a new way of building based on factory-made units that are installed and connected on site to create functioning buildings. The manufacture of the modules is a specialist and often bespoke activity to a particular supplier, but architects and other members of the design team should know how to satisfy the structural and building physics requirements of buildings constructed using load-bearing modules.

This gap in knowledge was the starting point for this book, which is aimed at providing sufficient information to understand and use the different forms of modular systems in building construction. In the course of this research, case studies were made using a range of residential, educational, and health sector buildings. This required close liaison with modular suppliers in both the UK and elsewhere in Europe, who provided much background information.

The new areas of application of modules are in highrise residential buildings and in specialist health sector buildings, including extensions to existing buildings. These applications highlight the key benefits of rapid and high-quality construction, and economy of scale in manufacture. Where the client is able to put a business value to these benefits, then modular construction is more likely to be the preferred choice.

The book brings together information on steel, concrete, and timber modules and describes their particular features and key design aspects. It draws on existing information from the Steel Construction Institute, the Concrete Centre, and Build Offsite in the UK, and refers to modern design standards and to the Eurocodes for these materials.

·The concepts and systems that are presented are not exhaustive, but it is hoped that this book will act as a key reference for designers wishing to use modular construction and will encourage a creative use of this new building technology. The book may also be used at an undergraduate and a postgraduate educational level, and also in continuous professional development.

Mark Lawson
Ray Ogden
Chris Goodier

Acknowledgements

This book was prepared by Professor R. Mark Lawson, The Steel Construction Institute, UK, and professor of construction systems in the Faculty of Engineering and Physical Sciences, University of Surrey; Professor Ray G. Ogden of the Oxford Institute of Sustainable Development, Oxford Brookes University; and Dr. Chris I. Goodier of the School of Civil and Building Engineering, Loughborough University. Nick Walliman of Oxford Brookes University prepared the background information and text on educational and medical buildings.

It draws on information provided by the Steel Construction Institute and companies active in the modular industry. Additional photographs and case study information were kindly provided by Caledonian Modular, Elements Europe, Futureform, Unite Modular Solutions, Elliot Group, Yorkon in the UK, and NEAPO in Finland.

Special thanks go to Maureen Williams, who typed the text, Sian Tempest, who prepared many of the drawings, and the teams at Oxford Brookes, Loughborough, and Surrey Universities, and Steel Construction Institute, who provided much of the background research information. Particular thanks go to the expert help of Nick Walliman, Chris Kendrick, Franco Cheung, and Nick Whitehouse of Oxford Brookes University. Rory Bergin of HTA Architects kindly supplied drawings of high-rise modular buildings. Steve Barrett of Futureform, Joanne Bridges of Bridges Communications, Mike Braband of the Design Buro (formerly of UNITE Modular Solutions), and Tiina Turpeinen of NEAPO provided information on a number of the case studies.

About the authors

Professor Mark Lawson is professor of construction systems at the University of Surrey, and consultant to the Steel Construction Institute (SCI). He is a chartered civil and structural engineer and member of the American Society of Civil Engineers (ASCE). His BSc (Eng) is from Imperial College, and his PhD was earned from the University of Salford in the field of stressed skin design of steel-framed buildings. His career started with Ove Arup & Partners, and then with the Construction Industry Research and Information Association. In 1987, Professor Lawson became research manager at the newly formed SCI and has authored over 40 SCI publications in the fields of steel, composite, and light steel construction, and also in sustainable building technologies. In 2011, he was awarded two prizes by the Institution of Civil Engineers for published papers, including the Howard Medal. He has been involved in many EU projects in the steel construction sector and has led major projects on sustainable design in steel and on modular construction systems.

Professor Ray Ogden has a BA (Hons) and Dip. Arch in architecture and a PhD in mechanical engineering. He has been involved in construction-related research and teaching since 1986, including work related to light steel, off-site, and modular construction, building envelope design, and low-carbon solutions. He is currently professor and associate dean for research and knowledge exchange in the Faculty of Technology, Design, and Environment at Oxford Brookes University, where he is also director of the Tata Centre for the Building Envelope. Professor Ogden was formerly deputy head of the School of Architecture at Oxford Brookes University, and has also worked for Arup Associates. He has authored seven books and numerous technical papers and reports, and has been responsible for a wide range of research and live demonstration projects.

Dr. Chris Goodier is a senior lecturer in the School of Civil and Building Engineering, Loughborough University, having worked previously for BRE and Laing Civil Engineering. He is a chartered builder with 20 years' experience in all aspects of construction, including research, contracting, and consultancy, and has published more than 130 papers, books, reports, and articles. He was the lead expert for construction on the UK government's Foresight Sustainable Energy Management and the Built Environment futures project, and also authored the Royal Institute of Chartered Surveyors (RICS) 2011 Future of UK Housebuilding. He recently chaired the 2013 British Council's International Conference on Sustainable Construction and leads a significant ongoing portfolio of government- and industry-funded research in the areas of concrete materials, off-site technology, sustainability, infrastructure, renewable energy, and construction futures.

Executive summary

The technology and application of modular construction are developing rapidly. Design using modular or three-dimensional (3D) elements of construction requires knowledge of modular production, installation, and interfaces to other building elements. This also extends to an understanding of the economics and client-related benefits that influence design decisions, which are covered in this book.

The book reviews the generic types of modular construction and presents examples of their application. Examples of steel, concrete, and timber modular construction are presented, including their implications on building design and construction. The structural action of groups of modules is presented in terms of resistance to vertical loading, stability, and robustness.

Dimensional and spatial planning is crucial to the success of modular construction projects in all sectors. To maximise building use and flexibility, modular units may be combined with planar elements or structural frames in hybrid construction. This new aspect of design is covered, and building formats for modular and hybrid construction projects are also presented.

Cladding, services, and building physics issues are addressed. Thermal and acoustic performance and fire safety should satisfy modern regulations, and design details for good performance in these areas are presented. Aspects of transport, tolerances, and installation are covered.

Over 40 case examples of building projects using modular construction are presented. Background information is also given to assist in design. A sustainability assessment of a typical modular project is also provided to assist in understanding the wider benefits of modular systems.

Chapter 1

Introduction to modular construction

Modular construction has established itself in many sectors of the building industry over the last 15 years. Historically, the main use of modular construction was in portable or temporary buildings, but this prefabricated construction technology using volumetric units is now used in a wide range of building types, from schools, hospitals, offices, and supermarkets to high-rise residential buildings. This demand has been driven by the off-site nature of the construction process, which leads to quantifiable economic and sustainability benefits.

In 1998, the UK government report *Re-thinking Construction* (Egan) called for a change in the thinking of clients and the supply chain toward a partnership and less adversarial relationship, which has encouraged a longer-term investment in manufacturing facilities and in developing new ways of building, which has led to an interest in the use of prefabricated construction technologies.

The term *modern method of construction* (MMC) was defined by its improvements in terms of the targets set by *Re-thinking Construction* and was characterised by a greater use of off-site manufacture (OSM). As a result, many clients saw the longer-term development of OSM as key to their strategic business activities in terms of speed of construction, improved quality, and reliability. Modular construction is probably the most well-developed OSM technology, and the majority (up to 70%) of the value of the construction work takes place in a manufacturing environment.

Modular construction uses three-dimensional or volumetric units that are prefabricated and are essentially fully finished in factory conditions, and are assembled on site to create complete buildings or major parts of buildings. Murray Grove in Hackney, north London, constructed in 1999, was the first major modular building to win architectural plaudits. The 5-storey building comprised 80 modules in an L-shaped plan form and was constructed with external access walkways and courtyard balconies, as shown in Figure 1.1. This building used standard-sized modules in an architecturally interesting way, which met the needs of the residents and the social housing provider, the Peabody Trust.

This new way of building using prefabricated modular units leads to many constructional and sustainability benefits. However, the investment in the manufacturing process and in fixed facilities in a particular location requires an economy of scale to drive the financial benefits that accrue. Modular construction therefore requires a discipline among all members of the design and construction team to maximise the repetitive use of manufactured components, and to optimise the integrated design, supply, delivery, installation, and commissioning process.

This publication addresses the design, manufacturing, and construction of a range of building types using modular units, and identifies the key features of the off-site manufacturing process, which will help to inform potential users of how to design buildings using this relatively new technology.

1.1 DEFINITIONS

The definitions of off-site construction are presented in detail by Buildoffsite (Gibb and Pendlebury, 2006b), which has produced a glossary of terms for this relatively new sector of the building industry. The key definitions relating to this publication are

- Modular construction—Three-dimensional or volumetric units that are generally fitted out in a factory and are delivered to the site as the main structural elements of the building.
- Planar construction—Two-dimensional panels, used mainly for walls, that can be prefinished with their insulation and boarding attached before delivery to the site.
- Hybrid construction—Mixed use of linear elements, panels, and modules to create a mixed-construction system.
- Cladding panels—Prefabricated façade elements that are attached to the building to form the completed building envelope.
- Pods—Nonstructural modular units, such as toilets and bathrooms, that are supported directly on the floors of the building.

Figure 1.1   Installation of modular units and completed building at Murray Grove, north London. (Courtesy of Yorkon and Cartwright Pickard architects.)

Modular construction is generally used to create cellular-type buildings, which consist of similar room-sized units of a size suitable for transportation. Partially or fully open-sided modules may be manufactured, in which two or more modules create larger spaces. Modular units may also be manufactured for higher-value components of the building, such as

- Bathrooms
- Lift and stair units
- Mechanical serviced units
- Prefabricated roofs, often incorporating services

To assist in understanding the various forms of off-site manufacture (OSM), four levels of the construction process are proposed in Table 1.1 (based on an illustration by Gibb (1999) and reproduced by Buildoffsite (2006b)). Level 0 represents entirely site-based construction, such as reinforced concrete or masonry. OSM level 1 introduces some prefabricated elements, such as roof trusses or precast concrete floor slabs. The majority of current construction processes involve a combination of level 0 and 1 components.

OSM level 2 consists of pre-manufactured linear or planar structural systems, such as timber and light steel framing systems. Structural steel frames provide the structural skeleton to which the other elements are attached, and are also considered to be level 2. OSM level 3 involves use of a high proportion of prefabricated elements, such as modular units, that may be combined with planar elements. Level 4 applies to complete building systems, which are procured from one source and are based on modular and other forms of prefabricated elements.

Another relevant definition is that of an *open building systems*, which is the name given to a range of building technologies that allow for interchange of components to create more flexible building forms. The International Council for Research and Innovation in Building and Construction (CIB) Working Group W104 is currently exploring at an international level the development and implementation of open building systems using prefabricated components, such as open-sided modules.

Examples of the various forms of structural and services components that may be defined by their level of off-site manufacture are illustrated in Table 1.2. In OSM levels 1 and 2, the proportion of prefabricated components is typically in the range of 10 to 25% of the overall build cost, whereas in level 3, this proportion increases to 30 to 50%, and in level 4 to more than 70%. This percentage use of OSM leads to an approximately proportionate reduction in overall construction time relative to level 0. Further savings in time on site by

Figure 1.1 (Continued) Installation of modular units and completed building at Murray Grove, north London. (Courtesy of Yorkon and Cartwright Pickard Architects.)

Table 1.1 Illustration of various levels of building technologies in the context of off-site construction

Level	Components	Description of technology
0	Materials	Basic materials for site-intensive construction, e.g., concrete, brickwork
1	Components	Manufactured components that are used as part of site-intensive building processes
2	Elemental or planar systems	Linear or 2D components in the form of assemblies of structural frames and wall panels
3	Volumetric systems	3D components in the form of modules used to create major parts of buildings, which may be combined with elemental systems
4	Complete building systems	Complete building systems, which comprise modular components, and are essentially fully finished before delivery to the site

Source: Adapted from Gibb., A.G.F., *Off-site Fabrication—Pre-Assembly, Pre-Fabrication, and Modularisation,* Whittles Publishing Services, Dunbeath, Scotland, 1999.

using higher levels of OSM may be achieved in projects that are likely to be built in poor weather or in difficult site working conditions.

Essentially, in modular and other off-site construction methods, slow unproductive site activities are replaced by more efficient and faster factory processes. However, the infrastructure of factory production requires greater investment in fixed manufacturing facilities, and also a repeatability of output to achieve an economy of scale in production. Also, the lead-in time to design and manufacture the prefabricated components is extended in relation to more conventional construction methods.

1.2 APPLICATIONS OF MODULAR CONSTRUCTION

In modular construction, the major parts of a building are produced fully finished in factory conditions rather

Table 1.2 Examples of levels of off-site manufacture (OSM)

Parameters	Levels of off-site manufacture (see Table 1.1)			
	1. Manufactured components	*2. Elemental or planar systems*	*3. Modular and mixed-construction systems*	*4. Complete building systems*
Examples of construction technologies	• Timber roof trusses • Precast concrete slabs • Composite cladding panels	• Structural steel frames • Timber framing • Light steel framing • Structurally insulated panels	• Prefabricated plant rooms • Modular lifts and stairs • Modules placed on podium level • Bathroom pods in framed buildings	• Fully modular buildings
Proportion of off-site manufacture (in value terms)	10–15%	15–25%	30–50%	60–70%
Reduction in construction time relative to level 0	10–15%	20–30%	30–40%	50–60%

Note: Levels of OSM based on work by Loughborough University (Gibb, 1999; Gibb and Isack, 2003). Level 0 represents site-intensive construction using traditional materials with little off-site manufacture, except for windows and doors, etc.

Figure 1.2 Student residences in Plymouth using modular construction. (Courtesy of Unite Modular Solutions.)

than on site. The benefits of modular construction may be focused on certain market sectors, where there is a demand for speed of construction, and economy in manufacture, or where reducing the disturbance of the building process is an important business or planning requirement.

The main applications of modular construction may be summarised as the following:

- Student residences, particularly medium- and high-rise buildings, such as in Figure 1.2
- Medium-rise residential buildings in urban locations, as in Figure 1.3
- Mixed residential and commercial buildings, as in Figure 1.4
- Private and social housing, as in Figure 1.5
- Hotels of 4 to 12 storeys, as in Figure 1.6

- Military accommodation, generally 3 to 4 storeys, as in Figure 1.7
- Health sector buildings, generally up to 3 storeys, as in Figure 1.8
- Educational sector buildings, generally up to 3 storeys, as in Figure 1.9
- Bathroom pods in hotel and offices, etc., as in Figure 1.10
- Secure accommodation and prisons
- Plant rooms and other serviced units, generally used in commercial buildings and hospitals
- Rooftop extensions to existing buildings
- New balconies and lifts attached to existing buildings

Table 1.3 summarises the construction sectors where off-site manufacture (OSM) is popular. The sectors where fully modular construction is used are mainly

Figure 1.3 Multistorey residential building in Manchester with retail outlets at the ground floor. (Courtesy of Yorkon.)

Figure 1.4 Residential building constructed on a podium structure with offices below on Commercial Road, east London. (Courtesy of Rollalong.)

those where OSM leads to tangible economic benefits. Other sectors that use modular construction are where a high level of service integration, specialist equipment, and off-site commissioning is required, such as in hospitals.

There are also increasing social pressures to build to higher densities in urban areas to meet the demand for single- or two-person and key worker accommodation. The use of modular construction in medium- to high-rise social housing projects has increased, particularly in inner cities, where the constraints on the construction process and site logistics often lead to the greater use of OSM. In mixed-use buildings, modular housing units may be supported by a structural podium in steel or concrete, which means that the street space below the residential levels can be used for offices, retail outlets, or car parking.

1.3 BENEFITS OF MODULAR CONSTRUCTION

The drivers for modular construction can be presented in terms of the well-understood decision-making

Figure 1.5 Town housing in Twickenham, west London, built in modular form. (Courtesy of Futureform.)

Figure 1.6 Hotel and high-rise residential development in Wembley, north London. (Courtesy of Donban Construction UK Ltd.)

Figure 1.7 Military accommodation in west London in fully modular form. (Courtesy of Caledonian Modular.)

Figure 1.8 Colchester Hospital constructed using modular construction. (Courtesy of Yorkon.)

parameters of cost, time, and quality, which can be quantified in financial terms. In modern building projects, there are planning and legal requirements to demonstrate sustainability in terms of its economic, environmental, and social impacts, which further extends the range of decision-making parameters.

The key advantages of modular construction in the context of cost, quality, and time may be summarised as the following:

- Shorter build times, leading to reduced site management costs and early return on the investment.
- Superior quality achieved by the factory-based construction process and predelivery checks.
- Economy of scale in production, particularly in larger projects or in repeated projects using the same modular specification.
- Excellent acoustic and thermal insulation and fire safety due to the double-skin nature of the

Figure 1.9 School buildings for the Harris Academy, Essex, built in modular construction. (Courtesy of Elliott Group Ltd.)

Figure 1.10 Installation of bathroom pods in a concrete-framed building. (Courtesy of Elements Europe.)

Table 1.3 Construction sectors most relevant to the use of off-site manufacturing (OSM)

Sectors for which OSM is most relevant	Levels of off-site manufacture (OSM)			
	2. Elemental or planar systems		3. Mixed-construction systems	4. Fully modular systems
	Structural frames	2D panels		
Housing		✓✓		✓
Apartments— multistorey	✓✓	✓✓	✓	✓✓
Student residences	✓	✓✓	✓	✓✓✓
Military accommodation				✓✓✓
Hotels	✓	✓	✓✓	✓✓✓
Office buildings	✓✓✓		✓	✓
Retail buildings	✓✓✓		✓	✓
Health sector buildings	✓✓✓	✓	✓	✓✓✓
Educational buildings	✓✓✓		✓	✓✓
Mixed use, e.g., retail/residential	✓✓	✓	✓✓✓	
Industrial, e.g., single storey	✓✓✓		✓	
Sports buildings	✓✓✓	✓	✓	✓
Prisons and security buildings	✓		✓	✓✓✓

Note: ✓✓✓, widely used; ✓✓, often used; ✓, sometimes used.

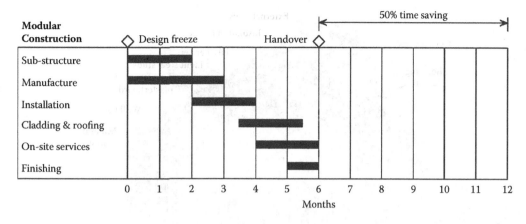

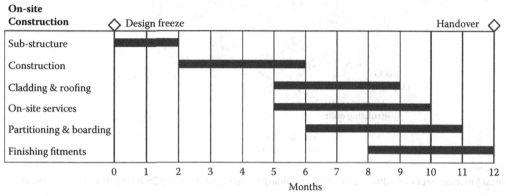

Figure 1.11 Relative construction periods for a 6-storey modular building compared to fully on-site construction.

construction, which means that each module is effectively isolated from its neighbours.

- Reduced design cost to the client (i.e., most of the detailed design work is carried out by the modular supplier).
- Lightweight, less material use, and less wastage compared to site-intensive construction, and greater opportunities for recycling in factory production.
- Increased productivity in factory production and reduced requirement for on-site labour. Installation of the modules is by specialist teams.
- Safer construction in terms of the factory and site activities.
- Less disturbance to the neighbourhood during construction, which is important where the adjacent buildings have to function without disruption.
- Ability to dismantle the building and maintain the asset value if the modules are reused elsewhere.

The economic benefits of modular construction are presented in more detail in Chapter 18. In the paper "Re-engineering through Pre-assembly," Gibb and Isack (2003) state that in order of importance based on a survey of clients, the perceived benefits of off-site manufacture were speed of construction, higher quality, lower cost, less wastage, and greater reliability.

The relative construction programmes for a 6-storey building in modular and on-site construction are presented in Figure 1.11. A 50% reduction in construction period is often achieved when using modular construction in comparison to fully on-site construction, depending on the building form and complexity. For buildings in which modules are placed on a podium level and where there are extensive on-site works, this saving in construction time can reduce to about 30%.

The excellent acoustic insulation provided by modular construction is another motivation to use it in residential applications. A typical junction between two modules side to side and above and below is illustrated in Figure 1.12, which shows the basic components of the modules.

1.4 HISTORY OF MODULAR CONSTRUCTION IN THE UK

Although modular units have been used for many years in portable buildings and as bathroom units in office buildings, designs using load-bearing modules only date from the early 1990s. One of the early examples was a student residence at Cardiff University designed by modular architect John Prewer; it is shown in

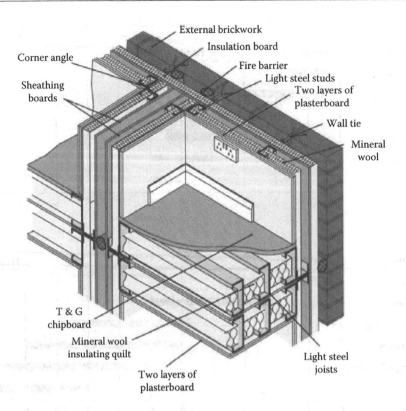

External brickwork

Insulation board

Corner angle

Fire barrier

Sheathing
boards

Light steel studs

Two layers of
plasterboard

Wall tie

Mineral
wool

T & G
chipboard

Mineral wool
insulating quilt

Two layers of
plasterboard

Light steel
joists

Figure 1.12 Typical junction between adjacent modules providing acoustic insulation. (Courtesy of the Steel Construction Institute.)

Figure 1.13 Student residence at Cardiff University using modular units circa 1990. (Courtesy of John Prewer.)

Figure 1.13. The award-winning Murray Grove project, completed in 1999, was the first modular building to gain public attention, and architect Cartwright Pickard was able to base its innovative design on the existing Yorkon modular system.

The Lillie Road project in Fulham, London, completed in 2003, used light steel framing, modular bathrooms, and a primary steel frame on the first floor to provide efficient use of space for this residential building. This completed building is illustrated in Figure 1.14.

Figure 1.14 Completed mixed-panel and modular project at Lillie Road, Fulham. (Courtesy of Feilden Clegg Bradley Studios.)

The Royal Northern College of Music's student residence in Manchester, completed in 2003, consists of 900 modules in a 6- to 9-storey courtyard configuration (see case studies), and was designed so that it could be dismantled and reused elsewhere on the campus. A mixed residential-retail development in Manchester for client OPAL consists of 1400 modules that are supported on a 2-storey steel-framed podium (see Figure 1.15). This project required the setting up of a temporary production facility only 5 miles from the project site in order to achieve "just in time" delivery of the modules.

Unite Modular Solutions (later called Lightspeed) set up its factory in the west of England to produce modules for the student residence and key worker sectors, and has completed over 50 major projects of up to 12 storeys high using fully modular construction. At its peak, the Unite factory produced study bedroom modules at a rate of up to 20 per day.

1.5 MODULAR CONSTRUCTION WORLDWIDE

One of the first examples of the use of large bathroom modules installed along with the primary structure was in Sir Norman Foster's Hong Kong Shanghai Bank in Hong Kong, which was built in the early 1980s. This also led to the wider use of modular serviced units in major commercial buildings in the London office boom

of the early 1990s. Modular housing is now widely used in the Far East, and notably is increasing in China and Korea.

1.5.1 Modular construction in Japan and Korea

Modular housing has been widely used in Japan since the early 1970s, and at the peak of output in the early 2000s, over 170,000 houses were sold per year, mainly to private purchasers. The main modular suppliers are Sekisui Heim, part of Sekisui Chemicals, Misawa, Daiwa, and Toyota Homes. Sekisui Heim produces modular houses from its six factories in various locations. Factories are highly automated and use an array of standardised components.

The marketing of modular housing in Japan is based on a high degree of user choice in the layout and fitments in the modules, and a fast design, manufacture, and installation turnaround. A house can be installed and finished in only 6 days, and therefore this is attractive in areas of Japan where land prices are very high. Modular construction is mainly used in Japan for 2- or 3-storey housing. An example of this type of modular housing is shown in Figure 1.16.

In Japanese systems, the modules are relatively small at 2.4 m wide and 3.6 to 5.4 m long, and are often built with open sides. Up to 12 modules make one large 2-storey family house. The modules often use a welded frame consisting of 100 mm steel box sections and 200 mm

Figure 1.15 Mixed commercial–residential development at Wilmslow Road, Manchester, using 1400 modules on a steel composite podium structure. (Courtesy of Rollalong and Ayrshire Steel Framing.)

Figure 1.16 Examples of modern Japanese modular housing. (Courtesy of Sekisui.)

deep edge beams. Modules are designed to be very resistant to earthquake effects, which is an important requirement in Japan. Cladding is often in the form of composite panels with preattached cladding or veneer in which the light steel profiles are embedded in the panel.

In the last 5 years, effort has gone into marketing zero utilities expense housing, and around 50,000 houses have been sold under this initiative. Misawa and Toyota have announced a joint venture on zero emissions modular housing, which includes inbuilt solar thermal and

Figure 1.17 Open-sided module used with an integral corridor used in a school building in Seoul, Korea. (Courtesy of POSCO.)

photovoltaic (PV) panels. More recently, Misawa has moved into modular social housing for rent.

In Korea, the steel company POSCO has developed a modular system for schools, which consists of a 12 m long by 3 m wide open-sided module with an integral corridor. In the example shown in Figure 1.17 for an elementary school in Seoul, the modules were supported on precast ground beams, and were installed in only 4 days. The same system has been used for military barracks.

1.5.2 Modular construction in North America

In North America, modular housing is based on the portable building industry, which is well developed at a regional level. Modules for housing can be very large (3.9 to 4.6 m wide by 12 to 15 m long) and are fully fitted out, clad, and delivered with a pitched roof, so that two modules form a large single-storey house. At the peak output in 2005, a total of over 40,000 modular houses was constructed in the United States, mainly in the northeastern states, and this represented 2.5% of the housing market at that time. Modular framing systems are generally based on timber, although light steel framing is used in some areas where wood is subject to termite attack.

In his book *Factory Design for Modular Home Building*, Mullen (2011) describes the types of timber-framed modules that are manufactured in the United

States, mainly for single-family houses. Approximately 44% of the houses purchased used standard house designs with little or no customisation, 59% houses comprised two modules, and 31% comprised three or four modules. The typical floor area of a module is 40 to 60 m², which is larger than in Europe, mainly because of fewer transport restrictions in the less urban areas of the United States.

McGraw-Hill Construction (2011) presented a survey of prefabrication and modularisation in construction that included architects, clients, contractors, and engineers. They reported that the largest areas of application in the United States were in healthcare buildings, in college buildings and dormitories, and in buildings for manufacturing industries.

More recently, the Modular Building Institute (www.modular.org) has been active in promoting modular construction in the United States, and there have been notable successes in use of modular technologies in apartment buildings, schools, and offices. One of the first uses of modular construction in a multistorey building was in a 4-storey apartment project in San Francisco that achieved a LEED Platinum award. This project consists of 23 single-module apartments and was completed in 3 months.

At Atlantic Yards, New York City, a 32-storey 30,000 m² residential building is underway that consists of a braced structural steel frame that supports the modular units on each floor. The steel frame is installed at the same time as the modules are placed,

Figure 1.18 High-rise modular building in north London. (Courtesy of Futureform.)

and this building will be the tallest modular building in the world. A total of 930 modules will create 350 apartments.

1.5.3 Modular construction in Europe

The main market for modular systems is in residential buildings in the UK and Scandinavia, and in the medical sector in Germany and the UK. There are many suppliers of modular buildings, and some are also active in the portable buildings sector. Most modular manufacturers use steel framing systems, but some use precast concrete and timber.

In the UK, it is estimated that at its peak in 2007, 8000 steel modules and up to 500 concrete modules were produced in various applications. The first companies active in modular production in the 1990s were Yorkon and Terrapin, who concentrated on the educational sector. Yorkon has since expanded its modular system into the medical and retail sectors.

In the last 10 years, the market in the UK has developed strongly in the areas of student residential

buildings, particularly in inner cities, hotels, and military accommodation. The use of modular construction has extended into high-rise buildings of 12 to 25 storeys, in which modules are clustered around a concrete core for stability. A recent example of a 16-storey residential building with a concrete core is shown in Figure 1.18.

The UK Government's SLAM and Aspire initiatives for military accommodation involved partnership agreements with various modular companies. Off-site manufacture has also been widely used by housing associations as a way of delivering higher quality and speed of construction.

In the modular precast concrete industry, the main applications are in hotels, military accommodation, and secure buildings such as prisons. These applications are covered in Chapter 3.

In the Nordic countries, there is strong incentive to use all types of prefabricated construction because of the short seasonal weather window for construction. In Finland, modular construction has traditionally been used as a secondary output from the modular cabin industry in shipbuilding. The steel company Ruukki has developed prefabricated cladding and enclosed balcony

Figure 1.19 Sheltered housing in Vantaa, Finland, built in modular construction. (Courtesy of NEAPO.)

Figure 1.20 Open-house modular system in southern Sweden. (Courtesy of Open House AB.)

systems, and also modular bathrooms, which are widely used both in new buildings and in renovation.

NEAPO, located near Tampere, Finland, has developed a double-skin steel panel system called Fixcel, which means that large modules can be manufactured up to 5 m wide and 16 m long without bracing. This system has been used in large residential projects and in rooftop extensions to existing buildings. An example of a 2-storey fully modular sheltered housing project is shown in Figure 1.19.

In Sweden, the Open House AB system was used in major housing projects in southern Sweden (see case studies). It is based on a 3.9 m grid of square hollow section steel posts, in which modules can be arranged and reorientated on this grid (see Chapter 2) A completed housing using this system is shown in Figure 1.20. IKEA has developed BokLok, which is a kit housing system using timber framing.

In the Netherlands, Spacebox was used for a 3-storey student residence in Delft and is shown in Figure 1.21. Flexline is also a modular housing system developed in response to the Dutch government's Industrialised, Flexible, De-mountable (IFD) initiative in the early 2000s.

In Germany, modular housing was provided by companies, such as Alho and Haller, but this has declined since the early 2000s. The Alho system used a timber frame, and was based on the concept of a generation house, in which owners could extend their houses as

Figure 1.21 Early example of student housing in Delft, the Netherlands, in modular form.

family sizes increased. Microcompact home (m-ch) is a concept designed for single-person living.

A large market for modular construction in Germany is in the health sector, where Cadolto and Draeger Medical provide highly serviced modules for all types of medical buildings. Modules can be large, subject to transportation, and are typically 4 m wide by 12 m long. They are manufactured using a welded steel framework so that open-sided modules can be created for specialist rooms, such as operating theatres (see Chapter 6).

In northern Spain, Modultec is a large modular company, which concentrates on educational and residential projects.

1.6 BACKGROUND STUDIES

In most developed countries, the built environment represents more than 40% of the total energy consumption of the country, and the European Commission, through its directive *Energy Performance of Buildings*, requires significant operational energy reductions in new buildings in order to reduce CO_2 emissions. The housing sector has been targeted as one of the main areas where important reductions in CO_2 emissions can be achieved. In the UK, this sector is responsible for 27% of the UK's total energy consumption. The main way in which improvements in new buildings can be achieved is through national building regulations in order to implement changes in practice that are intended to reduce energy loss through the building fabric.

The UK's Code for Sustainable Homes (Department for Communities and Local Government, 2010) is now mandatory in the housing and residential sector, and the Building Research Establishment's Environmental Assessment Method (BREEAM) is widely used in other sectors. Coupled with requirements for energy reduction and renewable energy provision in buildings of all types, off-site manufacture leads to a more reliable way of delivering these sustainability targets. Furthermore, the government's planning guidance PPG3 (ODPM, 2005) promoted mixed-use developments in urban locations and reuse of former industrial sites (brownfield sites). This has led to a demand for building technologies that are fast to construct, lightweight, and less site intensive. This planning guidance was superseded in 2011 by *Planning Policy Statement 3: Housing (PPS3)*.

In his report *Social and Economic Value of Construction*, Pearce (2004) identified the problems that the construction industry has in adapting to rapid demand and technology change. Although the UK construction industry has over 1.5 million participants, it is very diverse and lacks critical mass in many sectors, such as in off-site construction. Following this report, the number of companies active in off-site manufacture has increased. Pearce also noted that the construction industry consumes around 7 tonnes of building materials per person per year, and up to 20% of all materials are wasted at various stages of the construction process, which should be reduced by better design and efficient utilisation of materials.

Pearce (2004) identified various requirements of the construction industry to meet new challenges, which included

- Standardisation of building components
- Lightweight and stronger materials
- Wider user of information technology
- Wider use of off-site manufacture (OSM)
- Improved design for health and well-being
- Flexible use over time
- Integrated supply chains

These requirements are met in part by an expansion in the use of modular construction technologies. A UK government briefing paper, *Modern Methods of House Building* (POST, 2003), identified the need for a step change in the ability of the construction industry to meet demand for 3 million new homes by 2016. Due mainly to global economic factors, house building in the UK has fallen from around 180,000 units in 2007 to just over 100,000 units in 2012, and this has added to pressure on social and affordable housing. Apartments, generally in two- and three-person formats, account for about 35% of current house building in the UK. The proportion of the housing sector that is funded through public housing associations is about 20%, and it is this sector that has been the most receptive to off-site manufacture.

Also, as land prices have increased, the actual cost of the construction of modern houses is, in many areas, only around 30% of the sales price in the UK. For 2- or 3-storey private houses, the median build cost is £700 to £800/m^2, and for medium-rise apartments, the median construction cost is about £1200/m^2, depending on the location.

The supply side for off-site manufacture (OSM) in all materials, particularly timber framing and light steel framing, and in steel and concrete modular construction, has grown over the last 10 years, despite a weakening construction market since 2007. Companies who supply OSM technologies into the building sector are increasingly under pressure to reduce costs to compete against more traditional forms of construction.

David Gann's team at Imperial College produced many influential reports, including those comparing Japanese industrialised housing and the car industry (Gann, 1996), on flexibility and user choice (Gann et al. 1999), on overseas study tours to Japan (Barlow, 2001; Barlow and Osaki, 2005), the Netherlands, and Germany, and a review of the supply side in off-site manufacturing in the UK (Venables, 2003).

Considerable research into off-site manufacture, including modular construction, has been conducted at Loughborough University over the last 15 years by Alistair Gibb and his team (https://offsite.lboro.ac.uk). He wrote a widely referenced book on preassembly, prefabrication, and modularisation (Gibb, 1999).

Buildoffsite is an industry membership organisation for the promotion of off-site technologies and applications, and was established in 2005. Buildoffsite has produced a glossary of off-site terms (Gibb and Pendlebury, 2006a), a market value report (Goodier and Gibb, 2005b), and a set of off-site cameo case studies (Gibb and Pendlebury, 2006b).

A series of interactive toolkits was developed to enable clients, designers, and contractors to achieve the benefits of off-site and modular construction, by providing guidance on the overall concept and details. The standardisation and preassembly S&P Project Toolkit (Gibb and Pendlebury, 2003) supplied by the Construction Industry Research and Information Association (CIRIA) was the first, and was followed by IMMPREST, an interactive model for measuring preassembly and standardisation benefits across the construction supply chain (www.immprest.com). This was further improved into IMMPREST-LA, a design support tool for off-site and standardisation in construction. It is a cost and value comparison tool for off-site construction, and contains a checklist of considerations, in the form of an interactive spreadsheet.

Goodier and Gibb investigated the manufacture and installation of off-site products and systems, together with the resulting skills implications. This work, for CITB ConstructionSkills (Goodier et al., 2006) and British Precast (Goodier, 2008), highlighted the need for formal training and qualifications more widely in the sector.

The size of the off-site sector in the UK was estimated as £2.2 billion in 2004, and probably reached over £4 billion at its peak in 2007. The potential for expansion of the different off-site sectors was also investigated (Goodier and Gibb, 2005, 2007). Adaptable Futures (www.adaptablefutures.com) is a recent joint Loughborough University project that concentrates on building adaptability and the use of off-site construction. Ongoing work includes research into the off-site strategy of main contractors (Vernikos et al., 2013), use of building information management (BIM) in off-site construction (Vernikos et al., 2014), and the interfaces between modules (McCarney and Gibb, 2012).

Winch (2003) at Manchester University described how the process of lean manufacturing may be applied to the construction sector and the role of off-site manufacture. David Birkbeck and Andrew Scoones (2005) presented a well-illustrated review of 12 case studies of the use of off-site manufacturing technologies in their book *Prefabulous Homes*, as applied to various scales of house building.

A report by Goodier and Pan (2010) for the Royal Institute of Chartered Surveyors on the future of UK house building also highlighted the potential for off-site manufacture to satisfy the demands in the housing sector, which are increasingly influenced by sustainability requirements and reductions in operational energy use. The issue of new business models for off-site housing procurement was highlighted as particularly important (Pan and Goodier, 2012).

The Construction Industry Council (CIC) in the UK has recently issued an *Offsite Housing Review* (2013). It highlights the ability of off-site manufacture to achieve improved quality levels and more reliable energy efficiency performance than more traditional on-site construction. The increase in skilled factory-based employment was also highlighted as a result of greater off-site manufacture.

The report highlights the need to increase the social rented segment of housing supply through either local authorities or housing associations, and states that an output level of 45,000 to 75,000 housing units is required by 2020 to maintain pace with the increasing population. This is the area where the benefits of off-site manufacture are likely to be strongest because of the opportunities for standardisation and speed of delivery, particularly in the urban sector. One driver to the increased use of modular construction is the greater number of one-person households, which is expected to rise to about 40% of the total number of households by 2020.

In the McGraw-Hill Construction (2011) survey of prefabrication and modularisation in construction, it was reported that 24% of clients in the United States saw a reduction in project budgets of up to 5%, and 19% saw a 5 to 10% reduction, and 17% saw a 10 to 20% reduction in budgets when using modular construction. Importantly, 82% of respondents felt that a key driver was the improvement in productivity.

Internationally, the CIB has set up two working groups on open building systems (W104) and industrialisation in construction (W119), whose aim is to research and disseminate information on these technologies worldwide. The report *New Perspectives in Industrialisation in Construction* (CIB, 2010) provides the state of the art on existing experiences of prefabrication in all materials.

1.7 FUNCTIONAL REQUIREMENTS FOR MODULAR BUILDINGS

Functional considerations may be divided into two areas: performance and regulatory requirements, and those that are dependent on the use and architectural form of the building. These functional considerations in the context of modular construction are summarised in Table 1.4.

Structural, thermal, acoustic, and fire resistance requirements are part of the design and manufacture of the modules, and are therefore the responsibility of the

Table 1.4 Functional requirements for modular components

Functional consideration	Comment on modular construction
Plan form	Dependent on module size, the strategy for stability, and issues such as fire evacuation of the building. Additional braced cores are often required for taller buildings.
Circulation space	Means of access to the modules require design of corridors or external walkways, and braced stair and lift cores.
Cladding	Cladding may be in the form of ground-supported brickwork (up to 3 storeys high) or lightweight cladding. In both cases, the cladding is normally attached to the modules on site. The modules are designed as watertight insulated units.
Roofing	Roofs may be manufactured as modules, or using conventional roof trusses. Flat roofs are not normally recommended in modular construction unless provision is made for water runoff in the module design.
Thermal insulation	High levels of thermal insulation are generally provided within the modules, which can be supplemented by additional insulation on the outside of external walls.
Acoustic insulation	Double-layer walls, and combined floors and ceilings, provide excellent acoustic separation.
Fire safety	90 min fire resistance is generally achieved by the measures adopted for acoustic insulation. 120 min fire resistance is achieved by additional boards. Fire spread between the modules is prevented by use of fire stops.
Services distribution	Modules are generally manufactured as fully serviced units, and service connections are made externally to the modules. Corridors provide useful zones for service distribution.

modular supplier. However, the effective integration of modules into a complete building is more the responsibility of the client's design team, led by the architect. This should address issues such as the overall stability and robustness of the structure using modules, services distribution, attachment of cladding, access and circulation space, fire safety, etc.

The architecture of modular buildings is directly related to the use of similar three-dimensional components, which may accommodate some variation in size and layout, but are otherwise constrained by manufacturing and transportation requirements. Therefore, architectural design requires early dialogue with potential modular suppliers at the concept stage in the planning of these buildings. When the particular modular system has been chosen, the detailed design should be developed in close cooperation with the modular supplier.

The plan forms that are possible in modular construction are reviewed in Chapter 2. The servicing strategy is also linked to the particular plan form of the building, as although the modules are delivered as internally fully serviced, the horizontal and vertical routing of services through the building has to be considered carefully. In this respect, corridors often provide zones for horizontal service distribution and access for maintenance. This is considered in Chapter 15.

1.8 INTRODUCTION TO MATERIALS

1.8.1 Steel

Steel construction in its traditional form consists of skeletal frames, beams, and columns, and has established a track record in the multistorey commercial building sector. This form of steel construction uses hot-rolled I and H sections that are fabricated with their end connections, and are assembled on site using bolts, and sometimes by welding.

Steel-based modules use another form of steel in the form of galvanised steel strip that is cold rolled into C sections, in which the C sections are prefabricated into wall, floor, and ceiling panels, as shown in Figure 1.22. The C sections used in walls are 70 to 100 mm in depth and are in steel thicknesses of 1.2 to 2.4 mm, depending on their loading. These C sections are placed at 300 to 600 mm spacing to suit plasterboard dimensions. Floors use 150 or 200 mm deep sections in typically 1.5 mm thick steel, depending on their span. The technology of the use of steel in modular construction is described in Steel Construction Institute (SCI Publications) P272 (Lawson et al., 1999), P302 (Gorgolewski et al., 2001), and P348 (Lawson, 2007).

Galvanised strip steel is supplied to BS EN 10346 (British Standards Institution, 2009), and the total thickness of zinc used in the galvanised coating is equivalent to 275 g/m^2, or approximately 20 microns per side. The zinc oxidises sacrificially in the event of contact with water and air, and so the steel is protected even when scratched. On-site measurements (presented by SCI P262) have shown that the design life of the light steel components within the building envelope is over 100 years (Lawson et al., 2009).

Corner posts may be in the form of hot-rolled steel angle sections or square hollow sections (SHSs), depending on the particular modular system. Open-sided modules can be manufactured using steel edge beams (typically 300 to 400 mm deep) that span between corner posts. The design of steel modules is presented in more detail in Chapters 2 and 12.

1.8.2 Concrete

Precast concrete is a well-established and efficient manufacturing industry, and products range from hollow-core slabs to beams and columns in structural frames. Concrete modules can be manufactured in two ways—either from precast 2D wall, floor, and ceiling panels, or as 3D modular units, which are generally cast with an open base. Concrete modules are often used in high-security applications, as they are extremely resistant to damage.

The reinforced walls of the module are normally 125 mm thick, and the reinforced ceiling slab is normally around 150 mm thick. When the modules are

Figure 1.22 Manufacture of light steel wall panel. (Courtesy of BW Industries.)

manufactured with open bases, the ceiling of one module forms the floor of the one above, thereby saving on weight and structural depth. Innovations also include under-floor heating by pipes embedded in the ceiling slab. Design of concrete modules is presented in more detail in Chapters 3 and 13.

Concrete modular units are often manufactured so that two or three rooms are provided within one module to maximise efficiency. Modules may also be combined with other planar precast concrete units to create longer span areas. Toilet blocks and multifunctional rooms are also delivered as modular units. Core areas often use L- and T-shaped precast wall panels that are designed to provide the overall stability of the building.

1.8.3 Timber

Timber framing has been widely used in the residential sector since the 1960s and is widely used in the United States for modular housing. Historically, timber framing was also used in modular construction, particularly in temporary or relocatable buildings. The form of construction is based on prefabricated timber wall panels using nominally 89 × 38 mm wall studs with a top and bottom track of the same section. Normally the wall panels are sheathed with plywood or orientated strand board (OSB), and one or two layers of plasterboard are attached on the inside.

The floor and ceiling panels are manufactured with deeper joists (typically 225 mm deep), and in some systems the edge beams are manufactured as deep laminated beams, so that they span up to 10 m between corner posts. Timber modules can be designed up to about 4 storeys high. They have to be tied together at strong points, such as the corner posts, which are also used for lifting during installation.

1.9 ACCREDITATION OF MODULAR SYSTEMS

All construction products must conform to EU standards through European Technical Approvals (ETAs), for which the certifying body must be a member of the European Organisation for Technical Approvals (EOTA). For modular systems, this should ideally cover the manufactured product, which is the module itself rather than just its individual components. The materials used should also be CE (which means Conformité Européenne) marked, and is the case for building materials such as steel profiles and plasterboard. The ETA does not extend to the construction of the whole building, although module design should be as specified by the ETA. The ETA should cover the following:

- Description of the product, its intended use, and its characteristics
- Methods of verification, which cover the essential requirements of:
  - Mechanical resistance
  - Safety in case of fire
  - Hygiene, health, and the environment
  - Safety in use
  - Protection against noise
  - Energy, economy, and heat retention
  - Durability, serviceability, and identification
- Evaluation and attestation of conformity
- Assumptions of fitness for the purpose of the product

The Buildoffsite registration scheme, operated by Lloyd's Register, is a process-based assessment scheme designed to ensure that accredited organisations meet the benchmarked standards expected by clients who procure buildings from the off-site industry. (Lloyds Register, 2011). The scheme focuses on ensuring that accredited organisations have robust systems and procedures underpinned by a risk-based approach that enables them to competently, and safely, deliver products or services that meet the requirements of their clients. The categories of this accreditation include the following:

- Design
- Manufacturing
- Construction
- Project management

Accreditation lasts for 3 years and is subject to surveillance audits, which depend on the size of the organisation and the accredited scopes of work that are registered.

CASE STUDY 1: FIRST MAJOR MODULAR RESIDENTIAL BUILDING, LONDON

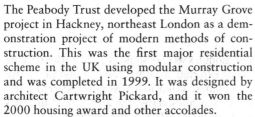

Roadside view showing the external walkways and X-bracing. (Courtesy of Yorkon and Cartwright Pickard Architects.)

Installation of a module by mobile crane (Courtesy of Yorkon.)

The Peabody Trust developed the Murray Grove project in Hackney, northeast London as a demonstration project of modern methods of construction. This was the first major residential scheme in the UK using modular construction and was completed in 1999. It was designed by architect Cartwright Pickard, and it won the 2000 housing award and other accolades.

The accommodation was targeted at key workers in London and for low-rental housing for couples. A high-quality architectural image was critical to the client in order to overcome the possible utilitarian perceptions associated with prefabricated buildings. The 5-storey L-shaped building occupies a corner site, and the cylindrical stair tower that encloses a glazed lift is located at the junction between the two wings. A private courtyard was created on the rear of the building, which is accessed via the secure entrance area.

The 74 Yorkon modules used light steel framing as their structure. A single-bedroom apartment comprises two 8 m long by 3.2 m wide by 3 m high modules, and a two-bedroom apartment comprises three modules. All bedrooms and living rooms had internal dimensions of 5.15 m by 3 m. The modules were manufactured with partially open sides so that the wider living and kitchen space crossed two modules.

In the architectural concept, internal corridors were omitted to save space, and access to the flats was provided by steel external walkways facing the street. The walkways are self-supporting and are X-braced by steel rods for stability, which also adds to the architectural effect. Each apartment also has a private balcony attached to the modules, which faces the communal garden on the rear.

The modular units were fully fitted out, serviced, and decorated and were fully equipped with bathroom and kitchen fittings, doors, and windows. Furthermore, the roof elements and the circular steel entrance, lift, and stairwell were delivered as modular steel elements.

The front elevations had a clip-on terra-cotta rain screen cladding system chosen for both its architectural qualities and its ability to be integrated into a dry construction system. The rails used to attach the tiles were pre-fixed to the modules. On the rear façade, cedar wood was chosen to give a softer feel to the courtyard. The balconies on the rear elevation were ground supported by a single tubular column and were tied into the modules at each floor level. Perforated aluminium screens form a translucent veil in front of the balconies and stair tower.

CASE STUDY 2: ROYAL NORTHERN COLLEGE OF MUSIC, MANCHESTER

View of the RNCM building from Booth Street West.

The Royal Northern College of Music (RNCM) is situated on Manchester's Oxford Road, and RNCM needed student accommodation close to the campus. In 2001, work began on a 9-storey student residence on the adjacent Booth Street West. The choice of modular construction was borne from the RNCM's desire to complete the project in 12 months to meet the start of the academic year, and also because there was a possibility that the building might be dismantled and moved to another location in the event of restructuring of the whole campus.

The building was designed in a square plan form around a central courtyard, and the building height ranges from 6 to 9 storeys. It was constructed over an underground car park with a concrete transfer slab at a semibasement level. Study bedroom modules were placed on either side of a central corridor, and the four corners of the building housed the stairs and lifts. They were constructed in steelwork and were braced to provide the stability of the building. The projecting roof was also supported by the upper modules.

Modules were manufactured by Caledonian Modular to a structural layout by their consultants, the Design Buro. The modules were manufactured with corner posts and steel edge beams but had light steel infill walls and floor joists.

The modules were designed to be weathertight by use of external sheathing boards and protective membranes. Wind loads were transferred laterally to the cores across the group of 32 modules per floor on each face of the building.

The rain screen cladding by Trespa was pre-attached to the modules, which meant that the time to scaffold and clad the building was eliminated. The joints between the modules were emphasised as part of the cladding design, but this required a high degree of accuracy in manufacture and placement of the modules.

The 612 study bedroom and 79 ancilliary modules were designed to a high level of soundproofing because of the need of the music students to practice in their rooms. The double-layer walls and floors in modular systems provided excellent acoustic insulation.

The installation of the modules and core steelwork from a concrete slab above an underground car park was completed in only 25 weeks. This meant that the building was completed in only 9 months, a savings of over 6 months on traditional building. The building was located on busy roads on three sides, and so the space for materials storage and site huts was limited. Module deliveries were also timed to miss the worst of the traffic. Modules were lifted straight from the lorry into position at the rate of 8 to 10 per day.

CASE STUDY 3: SOCIAL HOUSING, RAINES COURT, NORTH LONDON

View of the building. (Courtesy of Yorkon.)

View of a module being installed. (Courtesy of Yorkon.)

Raines Court, Stoke Newington in north London, was the Peabody Trust's second modular housing project, which demonstrated the ability of modular construction to offer architectural variety and to maximise the available space on the site. The 6-storey apartment block is T-shaped on plan in which the modules are configured to create a private courtyard with access walkways at the rear. The fully glazed entrance lobbies were built in steelwork.

Raines Court was commissioned, designed, and built by a partnership between architects Allford Hall Monaghan Morris, Wates Construction, and Yorkon. Installation of the 127 modules took place over a 4-week period. The contract period was only 50 weeks from start on site, saving 20 weeks relative to site-based building methods.

At the ground floor, the 3.8 m wide modules provide eight living/working units. Above are 5 storeys of two-bedroom apartments with a wing of three-bedroom family accommodation to the rear. Two modules create a two-bedroom apartment, and alternate units were manufactured with integral balconies. One module provided the living/dining/kitchen area and the other the bedrooms and generous bathroom.

The length of the modules varied from 9.6 to 11.6 m, and so each module provided a spacious 40 m^2 floor area. A 3.8 × 2 m balcony area was formed as part of the module. The modules were only 3 m high, allowing for a 600 mm combined floor-to-ceiling zone. The modules are designed to be self-supporting by their corner square hollow section posts. Stability of the 6-storey building was provided by the braced walls of modules, supplemented by X-bracing around the steel-framed access cores.

The façade to the main street was clad with lightweight shiplap-profiled zinc panels, with zinc cover strips to mask the joints between the modules. The panels were clipped onto a subframe directly attached to the modules in manufacture.

The courtyard elevations were finished with vertical larch timber cladding to add warmth to the finish of the external envelope. A glazed roof overhang on the sixth floor provides shelter for the access decks at the rear of the building. Square glass screens along the walkway provide further protection outside the entrance to each apartment.

CASE STUDY 4: MIXED MODULAR AND PANEL CONSTRUCTION, FULHAM

Courtyard view of the 6-storey building on Lillie Road. (Courtesy of Feilden Clegg Bradley.)

Internal view of X-braced cross-walls and modular bathrooms.

Light steel framing and modular construction was selected for the Peabody Trust's third major innovative housing project, in this case on Lillie Road, Fulham, west London. Specialist constructor the Forge Company and consulting engineer Michael Barclay Partnership, conceived a mixed-panel and modular structure, in which all the components were prefabricated. The project was completed in 2002.

The project was on the site of a former school, and for this inner city locality, reduced disruption due to the construction operation was an important client criterion in the choice of method of construction. It consists of 65 apartments, each of approximately 50 m² floor area, that were constructed in three blocks, the largest of which is 6 storeys high. The buildings are arranged around a sports arena that was built on belowground car parking. The construction period was reduced to 68 weeks, a savings of 20 weeks on the alternative in situ blockwork or concrete construction.

The 6-storey building is made from prefabricated light steel panels, floor cassettes, and bathroom modules that were all X-braced for stability.

The bathroom modules were also designed to be load bearing, so that their walls and floors contribute to the resistance to vertical and lateral loads. The preassembled floor cassettes used 200 mm deep C sections, and the wall elements used 100 mm deep C sections in 1.2 to 2.4 mm thickness, depending on the applied loads.

Architect Feilden Clegg Bradley continued the theme of prefabrication by choosing a lightweight stack-bonded terra-cotta tiling system as a rain screen façade. Aluminium rain screen cladding was used at higher levels. A sedum roof on the lower blocks reinforced the "green" landscape. Rectangular hollow section (RHS) members were introduced as expressed steelwork on the end façade, and also in the balconies. They were installed at the same time as the light steel wall panels.

The external walls achieved a U-value of 0.2 W/m²°C for a high level of energy efficiency by placing mineral wool between the C sections and also external to the wall. The separating floors and walls achieved an airborne sound reduction of over 63 dB, by using mineral wool and two layers of sound-resistant plasterboard.

CASE STUDY 5: MIXED-USE MODULAR BUILDING, MANCHESTER

View of the 8-storey mixed-use building from Wilmslow Road.

Partially completed modules built in the nearby field factory set up for this project.

The 8-storey building on Wilmslow Road, Manchester, for developer OPAL comprised a total of 1425 modules supported on a steel-framed podium structure at the first floor, which housed retail premises and a car park below-ground. The modular rooms are occupied by students of Manchester University, but also include some social housing.

The super-structure above the podium level was originally conceived in timber framing, but was replaced by a modular solution in steel. The modular suppliers, Rollalong, worked closely with their architects, Design Buro, to offer a design that could be completed in a narrow window from February to September 2002, in time for the intake of students.

The mixed residential-commercial development incorporates retail outlets, a health club, 130 key worker apartments (for rent), and six rooms for people with disabilities. The ground floor retail and basement car park levels were designed using a primary composite steel frame with a column and beam grid designed to support pairs of modules on each floor above.

Rollalong rented factory space in nearby Wythenshawe, and was able to set up a 10-line production of modules with an 8-day cycle of boarding, servicing, and fit-out before delivery to site. This is the first example in the UK of a field factory set up for one project.

A total of 945 study bedrooms using single modules, and communal areas using pairs of open-sided modules, were installed. A "man basket" system was used for installation, which was approved by the Health and Safety Executive (HSE). A peak installation rate of 28 modules a day was achieved by the nine-man team over the 4 months of the installation period on site.

The modules used the Ayrframe system, which comprises a grillage of C and top hat sections to create a stiff structure. Standard modules of 2.4 and 3.6 m width were arranged in three-, four-, and five-bedroom clusters around kitchens and communal areas. Corridors inbuilt within the modules reduced the site work and achieved weathertightness during construction. An integrated modular stair and lift shaft was also an important innovation.

The podium structure on which the modules were placed consists of 9 m long span I section beams acting together with a 170 mm deep composite slab on steel decking. The light weight of the 7 storeys of modules was an important factor in the design of the podium structure.

A rain screen cladding system was selected in order to achieve the rapid-build programme. It consisted of terra-cotta tiles on a substructure fixed through the cement particleboard facia to the modules. On the courtyard area, an aluminium rain screen cladding was used.

CASE STUDY 6: KEY WORKER HOUSING, WATERLOO, LONDON

View of the 3-storey building from Barons Place, Waterloo.

View of prefabricated steel walkways and stairs on the rear façade.

Keep London Working, an initiative by the Peabody Trust, estimated that 7000 new affordable homes for key workers are required in London every year. In 2005, Spaceover completed a demonstration project of affordable housing, called Barons Place, near Waterloo. The 3-storey building consists of 15 modules arranged in one- or two-bedroom configurations. The construction was completed to a tight cost schedule of £50,000 for a one-bedroom flat (excluding the land). On-site operations were managed by contractor Clancy Docwra.

Architects Proctor and Matthews designed efficient apartment layouts based on modules of 18 and 25 m² plan areas. A two-bedroom apartment of 54 m² comprised three modules, each of 18 m² plan area, and a one-bedroom apartment comprises two modules. The larger 25 m² module provided a one-bedroom studio apartment with integral kitchen and bathroom.

The 15 modules were installed over one weekend in order to minimise their impact on traffic in the Waterloo area. Modules were fully fitted out by Rollalong in Dorset and delivered to site with full-height glazed patio doors. The modules were up to 3.6 m wide and 7 m long. Partition walls can be positioned to meet a particular apartment layout.

A serviced zone was included in the corridor rather than in the module to allow for vertical service runs and interconnections. Modules are finished internally with plasterboard and with Trespa wall panels in the bathroom. All doors and bathrooms are suitable for disabled access.

The external walkways and balconies were positioned later and were prefabricated in galvanised steel C and tubular members. The roof to the walkway was in Kalzip cladding. The cladding in this project used lightweight cementitious panels, designed as a rain screen. The modules are fully weathertight in both the temporary and permanent conditions.

The servicing strategy includes an efficient electrical storage heating system, mechanically assisted Passivent for cooling, and other features such as dynamic insulation. Full-height patio doors were also used with featured balconies for cleaning. A U-value of 0.2 W/m²°C was achieved in the cladding design.

REFERENCES

Barlow, J., and Osaki, R. (2005). Building mass customised housing through innovation in the production system. *Environment and Planning*, 37(1), 9–21.

Birkbeck, D., and Scoones, A. (2005). *Prefabulous homes—The new house building agenda*. Constructing Excellence, London, UK.

British Standards Institution. (2009). *Continuously hot-dip coated steel flat products. Technical delivery conditions*. BS EN 10346.

Building Research Establishment Environmental Assessment Method (BREEAM). www.breeam.org.

CIB, International Council for Research and Innovation in Building Construction. (2010). *New perspectives in industrialisation in construction: A state of the art report*. CIB Report 329.

CIB, International Council for Research and Innovation in Building Construction. *Open building implementation*. W104. www.cibworld.nl.

CIB, International Council for Research and Innovation in Building Construction. *Customised industrial construction*. W119. www.cibworld.nl.

CLG (Communities and Local Government). (2011). *Planning policy statement 3: Housing (PPS3)*.

Construction Industry Council. (2013). *Offsite housing review*. London. www.cic.org.uk.

Construction Products Association. CE marking. www.constructionproducts.org.uk/sustainability/products/ce-marking.

Department for Communities and Local Government. (2010). *Code for sustainable homes—Technical guide*. London. www.gov.uk/government/publications.

Egan, J. (1998). *Re-thinking construction, the Report of the Construction Task Force* [the Egan report]. Office of the Deputy Prime Minister, London, UK.

Gann, D. (1996). Construction as a manufacturing process—Similarities and differences between industrialised housing and car production in Japan. *Construction Management and Economics*, 14, 437–450.

Gann, D., with Biffin, M., Connaughton, J., Dacey, T., Hill, A., Moseley, R., and Young, C. (1999). *Flexibility and choice in housing*. Policy Press, London.

Gibb, A.G.F. (1999). *Offsite fabrication—Pre-assembly, prefabrication and modularisation*. Whittles Publishing Services Dunbeath, Scotland.

Gibb, A.G.F., and Isack, F. (2003). Re-engineering through pre-assembly. *Building Research and Information*, 31(2), 146–160. doi: 10.1080/09613210302000, https://dspace.lboro.ac.uk/2134/9018.

Gibb, A.G.F., and Pendlebury, M.C. (2003). *Standardisation and pre-assembly—Project toolkit*. Report C593. Construction Industry Research and Information Association (CIRIA), London, UK.

Gibb, A.G.F., and Pendlebury, M.C. (2006a). Buildoffsite cameo case studies. www.Buildoffsite.org.

Gibb, A.G.F., and Pendlebury, M.C. (eds.). (2006b). Glossary of terms for offsite. Buildoffsite, London. www.Buildoffsite.org.

Goodier, C.I. (2008). Skills and training in the UK precast concrete manufacturing sector. *Construction Information Quarterly*, 10(1), 5–11. http://hdl.handle.net/2134/5463.

Goodier, C.I., Dainty, A.R.J., and Gibb, A.G. (2006). *Manufacture and installation of offsite products and MMC: Market forecast and skills implications*. Report for CITB ConstructionSkills. Loughborough University.

Goodier, C.I., and Gibb, A.G.F. (2005a). The offsite market in the UK—A new opportunity for precast? In Borghoff, M., Gottschalg, A., and Mehl, R (eds.), *Proceedings of the 18th BIBM International Congress*, Woerden, Netherlands, pp. 34–35. http://hdl.handle.net/2134/6013.

Goodier, C.I., and Gibb, A.G.F. (2005b). The value of the UK market for offsite. www.Buildoffsite.org.

Goodier, C.I., and Gibb, A.G.F. (2007). Future opportunities for offsite in the UK. *Construction Management and Economics*, 25(6), 585–595. http://hdl.handle.net/2134/3100.

Goodier, C.I., and Pan, W. (2010). *The future of UK housebuilding*. RICS, London. http://hdl.handle.net/2134/8225.

Gorgolewski, M., Grubb, P.J., and Lawson, R.M. (2001). *Modular construction using light steel framing: Residential buildings*. Steel Construction Institute P302.

Lawson, R.M. (2007). *Building design using modular construction*. Steel Construction Institute P348.

Lawson, R.M., Grubb, P.J., Prewer, J., and Trebilcock, P.J. (1999). *Modular construction using light steel framing: An architect's guide*. Steel Construction Institute P272.

Lawson, R.M., Way, A., and Popo-ola, S.O. (2009). *Durability of light steel framing in residential buildings*. Steel Construction Institute P262.

Leadership in Energy and Environmental Design (LEED). www.usgbc.org.

Lloyds Register. (2011). Buildoffsite registration scheme factsheet. Version 3. www.lloydsregister.co.uk/schemes/buildoffsite.

McCarney, M., and Gibb, A.G.F. (2012). *Interface management from an offsite construction*. In S.D. Smith (ed.), *Proc. ARCOM*, September. Edinburgh, UK, pp. 775–784.

McGraw-Hill Construction. (2011). *SmartMarket Report: Pre-fabrication and modularization in construction; increasing productivity in the construction industry*.

Modular Building Institute. (2012). *Permanent modular construction 2012 annual report*. Charlottesville, VA. www.modular.org.

Mullen, M.A. (2011). *Factory design for modular home building*. Constructability Press, Winter Park, FL.

ODPM (Office of the Deputy Prime Minister). (2005). *Planning policy guidance note 3*. Housing, London.

Pan, W., and Goodier, C.I. (2012). Housebuilding business models and offsite construction take-up. *Journal of Architectural Engineering*. doi: http://dx.doi.org/10.1061/(ASCE)AE.1943-5568.0000058, https://dspace.lboro.ac.uk/2134/9738.

Pearce, D. (2004). *Social and economic value of construction*. New Construction Research and Innovation Panel (nCRISP), London, UK.

POST. (2003). *Modern methods of house building*. Briefing paper for the Parliamentary Office of Science and Technology, London.

Venables, T., Barlow, J., Gann, D., Popa-Ola, S., et al. (2003). *Manufacturing excellence: UK capacity in offsite manufacture*. Housing Forum, UK.

Vernikos, V.K., Goodier, C.I., Robery, P.C., and Broyd, T.W. (2014). B.I.M. and its effect on offsite in civil engineering. *Institution of Civil Engineers Management Procurement and Law*, BIM special issue, April 2014 (forthcoming).

Vernikos, V.K., Nelson, R., Goodier, C.I., and Robery, P. (2013). Implementing an offsite construction strategy within a leading UK contractor. ARCOM Conference, Reading, UK, September, pp. 667–677.

Winch, G. (2003). Models of manufacturing and the construction process—The genesis of re-engineering construction. *Building Research and Information*, 31(2), 107–118. doi: 10.1080/09613210301995.

Chapter 2

Types of steel modules

Modular construction systems using steel components are built mainly using wall, floor, and ceiling panels that are manufactured from cold-formed galvanised steel C sections, supplemented often by corner posts in the form of hot-rolled steel angles or square hollow sections. The panels are formed into 3D modules and are boarded internally and usually sheathed externally, and then are fitted out and transported to the construction site.

Open-sided modules may be manufactured using corner posts and edge beams. The forms of light steel construction are presented in this chapter, and their structural design is presented in Chapter 12.

The architecture possibilities using steel modules were first presented in SCI P272 (Lawson et al., 1999). Guidance on design of modular buildings is also presented in SCI P302 (Gorgolewski et al., 2001).

2.1 BASIC FORMS OF LIGHT STEEL MODULES

Three generic forms of modular construction using light steel framing exist, which are reviewed in Steel Construction Institute, SCI P348 (Lawson, 2007):

- Continuously supported or four-sided modules, where vertical loads are transmitted through the walls (see Figure 2.1)
- Open-sided or corner-supported modules where vertical loads are transmitted through corner and intermediate posts (see Figure 2.2)
- Non-load-bearing modules, often called pods, that are supported by the floor or a separate structure (see Chapter 3)

These three forms of construction are used in different applications, depending on whether cellular space, such as bedrooms of a hotel, or open plan space is required. The modules act as the primary structure of the building, but the stability of the group of modules may be enhanced by other steel components or even a separate steel structure. Bathrooms or small plant rooms are often manufactured as non-load-bearing pods that are supported by the main structure of the building. Examples of these pods are presented in Chapter 4.

2.2 FOUR-SIDED MODULES

Continuously supported or four-sided modules are supported on their longitudinal sides, which bear on the walls of the modules below. The walls comprise 70 to 100 mm deep C section studs that are placed singly or in pairs at 600 mm centres, depending on the vertical load applied to the wall. The end walls are usually highly perforated by large windows at one end and a door and service riser at the other.

The floor and ceilings normally comprise C section joists placed at 400 mm centres that either are placed individually or are manufactured as part of a floor cassette with longitudinal edge members of the same depth. The assembly of these 2D panels into a three-dimensional modular unit is shown in Figure 2.3.

Corner columns, often in the form of hot-rolled steel angles or square hollow sections (SHSs), are used to provide local lifting points and attachments for other structural components, such as balconies. In some systems, the edge beams in the floor and ceiling provide the indirect means of load transfer by bearing on each other, but in most cases, the side walls provide direct load transfer. The indirect transfer method relies on the resistance of the edge beams to compression across their depth, and so this type of modular construction is limited to buildings of about 4 storeys in height. The direct method of load transfer through the side walls depends on the compression resistance of the C sections, which can be placed in pairs or rolled in thicker steel for higher loadings.

Reasonably large openings may be created in the walls of the modules, depending on the form of construction, as shown in Figure 2.2. In this case, the edge beam in the floor cassette is able to span up to about 2.5 m across the partially open sides of the module, depending on the building height.

Even in continuously supported modules, corner posts are used to provide lifting points, and the connection points to the other modules and structural elements.

Figure 2.1 Continuously supported module in light steel framing in which wall loads are transferred by bearing through the floor and ceiling cassettes. (Courtesy of Terrapin.)

Figure 2.2 Partially open-sided module with load-bearing walls. (Courtesy of PCKO Architects.)

For four-sided modules, hot-rolled steel angles or, in smaller modules, cold-formed steel angles (3 to 4 mm thick) are used as corner posts. Stability is provided either by placing X-bracing in the walls of the modules, or by diaphragm action of sheathing boards, or by a separate bracing system.

Figure 2.4 shows an adaptation of this modular technology in which a 12 m long module comprises an integral corridor. This avoids the need to construct the corridor in panel form and provides weather protection during construction. In this example, the module requires eight lifting points because the open-sided corridors are relatively flexible. This form of construction has been used by Caledonian Modular and Futureform in high-rise buildings (see case studies), where there are constructional benefits in the use of larger modules.

Figure 2.3 Assembly of a module from five panels, corner posts, and a floor cassette.

Figure 2.4 Corner-supported module with an intermediate corridor. (Courtesy of Kingspan Steel Building Solutions.)

2.3 CORNER-SUPPORTED MODULES

Corner-supported modules have posts at their corners, and sometimes at intermediate points, and edge beams span between the posts. In this way, the modules may be designed with open sides, although infill walls can be used to form the cellular space. The corner posts are generally in the form of square hollow sections (SHSs), and the edge beams may be parallel flange channels (PFCs) or heavier cold-formed sections.

Spans of edge beams are typically in the range of 6 to 12 m, and so they are typically 200 to 350 mm deep. It follows that the combined depth of the edge beams in the floor and ceiling is in the range of 450 to

Figure 2.5 Corner-supported module. (Courtesy of Kingspan Steel Building Solutions.)

750 mm, allowing for a 50 mm gap between the beams. A typical example of this type of module is shown in Figure 2.5.

Corner-supported modules have the advantage that they can be designed as open-sided between the posts. Intermediate posts may be introduced to reduce the span of the edge beams or for transportation purposes, as shown in Figure 2.6.

However, because the beam-to-post connections are relatively weak in terms of their bending resistance, the stability of the group of modules is provided by additional bracing that is often located around the stair and lift core. These types of open-sided modules are often used in the health and educational sectors (see Chapters 7 and 8).

2.4 OPEN-ENDED MODULES

Open-ended modules can be manufactured using a welded steel end frame. A typical example of a welded end frame using 250×150 mm rectangular hollow sections (RHSs) is illustrated in Figure 2.7. In this way, full-height glazing can be provided, and modules can also be combined along their length. This type of rigid end frame can be used to provide lateral resistance to wind loads acting on the long side of the modules for buildings up to 6 storeys high, depending on the number of modules in a horizontal group. An example of

this principle used in a hotel system in central London is shown in Figure 2.8.

2.5 HYBRID MODULAR AND PANEL SYSTEMS

In hybrid or mixed modular and panel systems, the modular units are used for the higher-value serviced areas, such as bathrooms and kitchens. The load-bearing wall panels and floor cassettes create the more flexible open space. This system was used in a demonstration building for Tata Steel, which is shown in diagrammatic form in Figure 2.9. In this project, the bathrooms, kitchens, and stairs were manufactured as a single 3 m wide and 1 1m long module that was shared between two apartments. The module supported prefabricated floor cassettes of up to 5.7 m span. This type of hybrid modular construction is explored in Chapter 8.

2.6 HYBRID MODULAR, PANEL, AND PRIMARY STEEL FRAME SYSTEMS

Modular construction is mainly used for medium-rise buildings of cellular form. Greater flexibility in building height and in internal planning can be achieved by the use of modules together with a primary steel structure. Various forms of mixed construction may be designed, as follows:

Figure 2.6 Corner-supported module with intermediate posts. (Courtesy of BW Industries.)

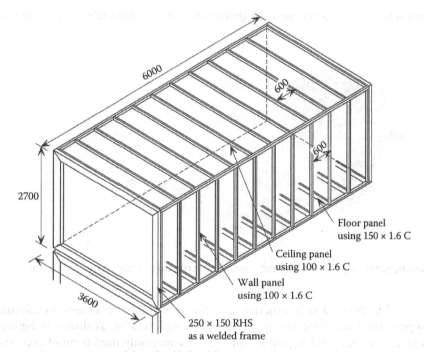

Isometric view module and welded end frame

Figure 2.7 Creation of flexible space using open-ended modules.

- A podium structure of typically one or two floors that is designed to support the modules above. The columns are placed at two or three times the width of a module, i.e., at 6 to 9 m typically.
- A framed steel or concrete structure provides the open plan areas on a particular floor, and the stacked modules provide the highly serviced areas or cores.

- A framed structure, in which non-load-bearing modules and wall panels are supported on the beams or concrete floor.

A podium structure is often used to provide commercial or communal space at the ground floor and car parking in the basement. Composite steel and concrete

Figure 2.8 Citizen M hotel in Southwark, London, with fully glazed open-ended modules. (Courtesy of Futureform.)

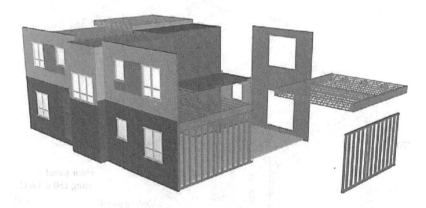

Figure 2.9 Demonstration building using mixed-panel and modular construction. (Courtesy of Tata Steel.)

construction may be used to create a stiff structure at podium level to support the load from the modules above. The podium structure may be designed to support typically 4 to 6 storeys of modules above. This approach is described in more detail in Chapter 10.

In the second approach, a framed structure may be designed in the form of slim floor beams in which the modules and floor cassettes are supported on the extended bottom flange of the beams so that they occupy the same depth as the floor. A pair of modules may be located within the column grid, and the corners of the modules are recessed in order that they fit around

the SHS or narrow H section columns in order to minimise wall widths, as shown in Figure 2.10.

A commonly used form of construction for multistorey buildings is to design a primary steel frame in composite construction or in slim floor construction and to use non-load-bearing light steel infill walls for external and separating walls. Bathroom pods may be slid into place, and in order to obtain a consistent level, their floor depth is the same as the built-up acoustic layers on the slab. The various types of pods are reviewed in Chapter 4.

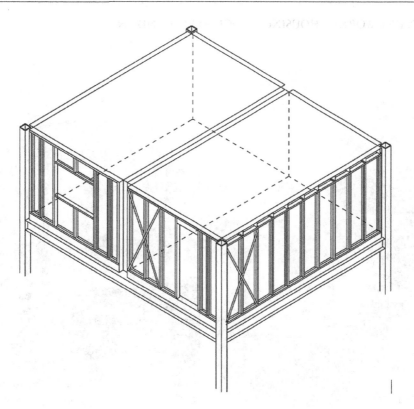

Figure 2.10 Recessed modules supported by primary steel frame.

CASE STUDY 7: KEY WORKER HOUSING, STOCKWELL, LONDON

Completed building viewed from Larkhall Lane, Stockwell.

Installation of modules along the front façade to the building.

Architect PCKO designed this 4-storey building in Stockwell, South London, for the Hyde Housing Association using light steel modules imported by manufacturer BUMA in Poland. The building consisted of four two-bedroom and four single-bedroom apartments. The two-bedroom apartments comprised two partially open-sided modules. A central stair module was installed at the same time, and the whole installation process took only 3 weeks. The construction was completed in 8 weeks from foundation level, and the main contractor, Rok, estimated that this represented only 20% of the construction time of brick- and blockwork.

Completed in 2004, the project value of £700,000 was cost-effective for this eight-apartment project. The client also wished to have the opportunity to dismantle and reuse the modules elsewhere. All cladding was attached to the modules in the factory. Steel balconies were ground supported. Apartments are electrically heated, which was economic for a building of this type with a highly insulated façade.

The modules are approximately 3.3 m wide and 9 to 11 m long. Two modules formed one apartment with either one or two bedrooms, depending on the length of the module. The modules were also designed with partially open sides to provide more useable living space. The 2.5 m wide stairs were also prefabricated as modules. All 20 modules were installed in less than 10 days, and finishing work was completed in the following 5 weeks.

Cladding consisted of both metallic panels and insulated render. Prefabricated steel balconies were later installed on the front and rear elevations to all apartments and were accessed through tilt and openable full-height windows built into the modules. The upper modules were provided with a shallow-pitch steel roof to further speed up the construction process.

Foundations were simple strip footings, and the lightweight steel structure, cladding, and roof reduced loads by over 50% in comparison to a reinforced concrete frame, which was considered at the early stages of the design. Materials, deliveries, waste, and number of site personnel were also reduced dramatically by the modular construction process.

CASE STUDY 8: CONFERENCE CENTRE—LEAMINGTON SPA

View of the large metallic cladding panels to the modular accommodation building.

Detail of attachment of rain screen panels to vertical runners fixed to the modules.

The Ashorne Hill Conference Centre near Leamington Spa required high-quality accommodation for its residential courses and turned to Terrapin for the design and construction to meet the tight time period for construction. This project gave the opportunity to work with Corus Living Solutions who manufactured, fitted out, and delivered the 27 modular bedrooms and plant room for this 2-storey residential building. The building was designed to complement the existing grade II listed mansion house and was completed in 2005.

The modules are 3.8 × 6.3 m in external plan dimensions, which included a bathroom. The modules are arranged either side of a 1.2 m wide central corridor with a large staircase at one end. Service connections were made in a vertical riser between pairs of modules.

The room-sized modules used 100 × 1.6 mm C sections for the walls and 165 mm deep C sections for the floor joists. The walls used fire-resistant plasterboard and large Fermacell panels internally. On the external walls, closed-cell insulation was directly fixed to moisture-resistant plasterboard. The floors comprise oriented strand board (OSB) on 19 mm plasterboard panels below 22 mm chipboard with polystyrene blocks on mineral wool placed between the floor joists for acoustic insulation. The total depth of the floor and the ceiling was 450 mm.

The external cladding is made from coated steel panels manufactured in the colour Orion from Tata Colors' Celestia range. These rain screen panels were prefabricated in sizes to match the window pattern and are supported on vertical rails that were preattached to the modules. The client chose this metallic finish to blend in with the traditional grey slate of the main building. The steel cassette panels of up to 2.3 m length and 1 m width were attached by nylon pins to the rails to form a rain screen with hidden fixings.

The overall construction period took only 5 months from start on site, and importantly, the 28-room modules were installed in only 3 days to create a weathertight enclosure for later fit-out and finishing. The building was designed to high standards of energy efficiency and comfort and to a high level of acoustic insulation.

The 3.83 m wide by 10 m long stair modules were supplied as open-topped, and the flight of stairs was supported by a crossbeam constructed as part of the top of the module. The floor of the upper module formed the stair landing. The V-shaped roof used internal guttering in downpipes located in the service zone between pairs of modules.

CASE STUDY 9: MODERN MODULAR HOTEL, SOUTHWARK, LONDON

Completed hotel on Lavington Street, Southwark.

View inside the atrium of the hotel.

Modular construction is widely used for hotels, where speed of construction leads to financial benefits to the hotel operator. A new hotel for Citizen M on busy Southwark Street on London's south bank has taken modular construction to a new level of design sophistication. Futureform's modular solution used a new form of 95 double-room and corridor modules of 2.5 m width and 15 m length, together with a small number of single-room modules.

The 6-storey hotel comprises 192 bedrooms, with the hotel reception, restaurant, and other facilities located at the ground floor. The modules are built to a high specification, including full-height windows, local heating and cooling of the air supply, mood lighting, and a high level of acoustic separation between the rooms. The modules are supported by a single-storey steel frame, which houses the hotel reception and restaurant at the ground floor.

It was designed for 90 min fire resistance with inbuilt high-mist sprinklers, which is in excess of Building Regulation requirements. The overall wall zone was only 200 mm, and the combined floor and ceiling zone was only 400 mm, which is remarkably narrow for a double-layer construction. A fully glazed façade wall was created by a welded frame using 80 × 40 RHS sections. This rigid frame provides the resistance to horizontal loads acting on the 5-storey assembly of modules, and also provides the attachment points between the modules.

The steel structure at the ground floor level is based on a 6 × 5 m grid, and each beam supports two modules. Steel columns aligned with the width of pairs of modules. Modules weighing up to 10 tonnes were lifted into place at an average rate of 6 per day by the 50 m boom of a 500-tonne mobile crane positioned roadside. The loads on the foundations were minimised by the use of lightweight modular construction. The stair and lift modular cores on the upper levels were installed at the same time as the bedroom modules.

The modular bedrooms were fully finished before delivery to site, and the corridors were finished by second fix services on site. Futureform's subcontract, as the modular designer and supplier, was only 35% of the total build cost of £14 million. The installation of the modules took only 5 weeks out of a 9-month construction programme, saving an estimated 6 months relative to more traditional building, which was very important to the client in the run-up to the 2012 Olympic Games. The project achieved BREEAM "Very Good." Measured airtightness of the modules was 5 m³/m²/h, which is significantly better than in traditional building. The measured sound reduction between rooms was over 60 dB.

REFERENCES

Gorgolewski, M., Grubb, P.J., and Lawson, R.M. (2001). *Modular construction using light steel framing: Residential buildings.* Steel Construction Institute P302.

Lawson, R.M., Grubb, P.J., Prewer, J., and Trebilcock, P.J. (1999). *Modular construction using light steel framing: An architect's guide.* Steel Construction Institute P272.

Lawson, R.M. (2007). *Building design using modular construction.* Steel Construction Institute P348.

Chapter 3

Precast concrete modules

Precast concrete elements are widely used in modern construction, and include planar elements, such as slabs and walls, and linear elements, such as beams and columns. These elements can be combined to form volumetric units, either as part of the casting process in a factory or assembled on the construction site. This chapter concentrates on the types of modular units that are cast monolithically in a factory. Guidance on structural design of concrete modules is presented in Chapter 13.

3.1 BENEFITS OF PRECAST CONCRETE MODULES

The particular benefits relating to the use of concrete modules and their on-site finishing activities arise from

- Formwork is efficiently used for production of concrete units in the factory.
- Installation rates of 6 to 10 modules per day can be achieved, taking account of the crane capacity and the distance over which the modules are lifted.
- Flush and continuous walls and ceilings are provided.
- Internal walls are cast integrally within the modules.
- The concrete walls do not usually require a full coat of plaster or other finish. A plaster skim coat is usually all that is needed on site.
- Service voids and electrical conduits can be built into the concrete.
- External concrete panels can be finished with a variety of surface treatments.
- Modules can be combined with planar walls and floors by forming recesses and bearing surfaces within the modules.
- Higher construction tolerances are achieved than in on-site construction.

Precast concrete elements reduce the requirements for formwork and scaffolding, thus saving costs through reduced on-site resources, as well as by shortening the on-site construction programmes. Most manufacturers of precast concrete modules provide the detailed design, delivery, and on-site installation. An example of a large concrete module being installed is shown in Figure 3.1.

Concrete construction has inherent benefits in terms of its fire resistance, sound insulation, and thermal capacity. The relatively high weight of concrete buildings means that strict vibration criteria can be met for specialist applications, such as laboratories and hospital operating theatres.

Precast concrete elements achieve higher accuracy and quality than in situ concrete. Precast concrete factories use in-house concrete production, thus ensuring consistency of supply and control of materials, which leads to greater reliability of colour, texture, and performance. A wide variety of high-quality finishes can be achieved.

The majority of precast concrete is manufactured within a day's travelling distance to where it is used. Control of materials and efficient factory processes also minimises wastage. Common with other modular systems, off-site manufacture reduces the level of activity on site, which can enhance the overall safety of concrete construction by eliminating labour-intensive formwork installation and striking, materials handling, etc.

3.2 PRECAST CONCRETE BUILDING FORMS

There are several forms of building construction that use precast concrete elements, as follows:

- Frame and floor slab
- Cross-wall construction
- Twin-wall construction
- Tunnel form construction
- Modular construction

Precast concrete elements may also be combined with in situ concrete, such as a concrete topping placed on precast floor units. Precast concrete frames are used for single-storey industrial buildings, offices, car parks, and some public buildings. Large precast concrete cladding units are also widely used in multistorey office buildings.

Figure 3.1 Installation of precast concrete modules on site. (Courtesy of Oldcastle Precast.)

Figure 3.2 Typical cross-wall construction. (Courtesy of Precast Structures.)

3.2.1 Frame construction

Precast concrete frames are mainly used for single-storey industrial buildings, car parks, and low-rise office buildings. The structural form consists of beams, columns, floors, shear walls, and specialist components, such as staircases.

3.2.2 Cross-wall construction

Cross-wall is an effective method of construction that uses precast planar components. Load-bearing cross-walls provide the vertical support and lateral stability to the building, as illustrated in Figure 3.2. Longitudinal stability is achieved by external wall panels or by diaphragm action of the floors, which transfer horizontal loads to the lift and stair cores.

3.2.3 Twin-wall construction

Twin-wall construction is a combination of precast and in situ concrete construction. Each wall panel consists of two precast reinforced concrete skins, which are held together by lattice reinforcement, as shown in Figure 3.3. The concrete skins act as permanent formwork, and they act structurally with the in situ concrete that is infilled between the skins. The weight of a panel is therefore reduced compared to a similarly sized fully precast panel, permitting the use of larger panels or requiring smaller cranes for installation. An example

of twin-wall construction of a multistorey residential building is shown in Figure 3.4.

3.2.4 Tunnel form construction

Tunnel form is a type of formwork system used to form cellular structures (Brooker and Hennessy, 2008). The system consists of inverted L-shaped half-tunnel forms that, when fitted together, form the full tunnel (see Figure 3.5). The system also incorporates gable end platforms and stripping platforms for circulation and to facilitate striking of the formwork. The cellular structure is formed by pouring the walls and slab monolithically in one pour, often at a rate of one floor per day.

A key issue to consider for a tunnel form project is whether the formwork can be lifted clear of the building and moved to its next position. Clearances of 5 m are generally required on at least one side of the building, although shorter tunnel form units can be used (but these will slow productivity).

3.3 MODULAR CONSTRUCTION IN CONCRETE

Hotels, prisons, and secure accommodation are the most common applications of modular precast concrete construction, as economy of scale in manufacture can be achieved. Modular precast concrete units can weigh up to 40 tonnes, although 20 tonnes is more typical. They are transported to site and craned into position onto a pre-prepared ground floor slab. Examples of modular construction in concrete follow.

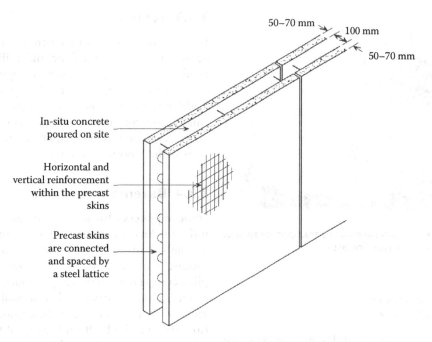

In-situ concrete poured on site

Horizontal and vertical reinforcement within the precast skins

Precast skins are connected and spaced by a steel lattice

50–70 mm
100 mm
50–70 mm

Figure 3.3 Typical twin-wall concrete panel.

Figure 3.4 Typical twin-wall project. (Courtesy of John Doyle Construction Ltd.)

Figure 3.5 Typical tunnel form formwork. (Courtesy of Outinord International Ltd.)

3.3.1 Hotels

Corridor-type construction can be achieved by the repetitive use of modular precast units. For hotels, the rooms can be finished by painting the internal concrete surface. Modules may also be manufactured with suitable cladding finishes, as shown in Figure 3.6. Corridors can be manufactured as planar elements, or as extensions to the room modules. If required, the modules can come with prefitted air conditioning units, furniture, additional external insulation, and external cladding.

Similar forms and sizes of modular precast units may also be used for student accommodation, military barracks, and key worker accommodation.

Figure 3.6 Precast modular hotel units being transported to site. (Courtesy of Oldcastle Precast.)

3.3.2 Prisons and secure accommodations

Prison cell blocks often use modular precast concrete construction. The walls and roof are cast in a single concrete pour using special moulds that are specifically designed to simplify the de-moulding process (see Figure 3.7). The base of the module is generally left open, so that the roof of the module below forms the floor slab. In this way, the concreting process is simplified and a single-layer floor is created. Window grilles and door openings can be cast into the reinforced concrete walls.

Figure 3.7 Casting a modular concrete unit. (Courtesy of Tarmac Precast Ltd.)

3.3.3 Schools

Precast concrete modules can be manufactured with open sides by using a floor and ceiling structure with rigid connections, as shown in Figure 3.8. This form construction is mainly used in single-storey school buildings, in this case with a bonded brick façade. A ribbed concrete roof slab can achieve spans of up to 12 m. The floor slab is supported on ground beams at intermediate points, and so is thinner than the roof slab.

3.3.4 Basement modules

Some suppliers also produce modular precast concrete units for basements up to 6 × 3.6 m on plan with flat or shaped inverts, which look similar to traditional precast concrete culvert units. Units can also be supplied complete with openings, doors, ventilation shafts, services, and stairways. The modular basement units are designed to enable a traditional masonry structure above to be built on the top of the basement unit, which also acts as the building foundation.

3.3.5 Bathroom pods

Bathroom pods can be manufactured in precast concrete, in which the structure consists of thin concrete walls and a floor that are reinforced with a single layer of steel mesh. Electrical conduits and pipework can also be cast into the concrete. To make efficient use of shared service risers, bathroom pods are usually located back-to-back around the service riser, and consequently up to four pods may be concentrated around one area of the slab. The weight of a concrete bathroom pod can be up to 4 tonnes.

Pods may be manufactured with or without a floor, depending on the adjacent floor level, as the finished floor surface of the pod should match the surface level of the surrounding floor. In some situations, it is acceptable to have a step from the general floor level into the bathroom, but in most cases, a thinner floor is required under the bathroom pod, or a screed is placed over the adjacent floor to bring it up to the same level as the floor of the pod. An example of a concrete bathroom pod is shown in Figure 3.9.

3.3.6 Precast concrete cores

Precast concrete cores may be used in any type of modular building or in open-plan framed construction, as illustrated in the case studies. Precast cores generally come with attachments for stairs and lifts. Stairs may be installed sequentially along with the modules on a given floor, but cores can be installed one or two floors in advance. Precast cores can also be manufactured to incorporate a pair of lifts.

Figure 3.8 Installation of precast modular units for a school. (Courtesy of Oldcastle Precast.)

Figure 3.9 Concrete bathroom pod being lifted into student residence built in cross-wall construction at the University of West of England (UWE), Bristol. (Courtesy of Buchan, Concrete Centre, 2007.)

CASE STUDY 10: MODULAR PRECAST CONCRETE CORES, HARTLEPOOL COLLEGE

Precast concrete lift modules during installation. (Courtesy of PCE Design & Build.)

Double-lift shaft module being lifted into place.

Precast concrete modular systems provide a solution for installing lift shafts quickly, accurately, and economically when compared to traditional in situ concrete construction. They can be used for a variety of buildings, ranging from educational and custodial to retail and residential buildings. Single- or multiple-lift shafts can be completed on site in as little as 1 day. The precast lift units are delivered to site on a "just in time" schedule and installed directly from the lorry. The lift core installation may be more than 5 storeys in advance of the rest of the building as the assembly of modules is stable up to 20 m high.

Main contractor Miller Construction installed a series of precast modular lift shafts for the core of the new Stockton Street Campus building of Hartlepool College in Teesside, which opened in 2012. The lift shaft design and construction was carried out by PCE Design & Build. The construction programme was dependent on the need to install the lifts and stairs rapidly, and five precast lift shafts ranging from 14 to 21 m in height were installed in just 5 working days.

The modular lift shafts comprising 27 single units and 10 double-lift sections were placed using a 100-tonne mobile crane. The larger lift modules were 4 m wide and 3 m deep, and the walls were 150 mm thick The segmental height of the units was up to 2 m, and so two units created a 1-storey height. The base sections were placed on prelaid plinths for precise verticality. Three single and one double core were completed in sequence, which were then capped at their tops to provide weather protection. The lifts were then installed inside the lift shafts, which also incorporated pockets for the crossbeams to support the lifts. This project also included 27 precast stair and 14 precast concrete landing units.

The modular units rely mainly on frictional bearing of one module on another, but their positional accuracy is achieved by connecting bolts between the units. The weight of a typical modular lift unit is 8 to 12 tonnes. The lift units are reinforced to resist the vertical loads applied to it from the weight of the lift. All the attachment points for lift guide rails can be installed in the factory.

REFERENCES

Brooker, O., and Hennessy, R. (2008). *Residential cellular concrete buildings: A guide for the design and specification of concrete buildings using tunnel form, cross-wall or twin-wall systems*. CCIP-032. Concrete Centre, London, UK.

Concrete Centre. (2007). *Precast concrete in buildings*. Report TCC/03/31. London, UK.

Concrete Centre. (2009a). *Precast concrete in civil engineering*. London, UK.

Concrete Centre. (2009b). *Design of hybrid concrete buildings*. London, UK.

Elliott, K.S. (2002). *Precast concrete structures*. BH Publications, Poole, UK.

Narayanan, R.S. (2007). *Precast Eurocode 2: Design manual*. CCIP-014. British Precast Concrete Federation, Leicester, UK.

Chapter 4

Other types of modules

This chapter reviews the use of other types of modules, which includes reuse of shipping containers, manufacture of small bathroom pods, larger plant rooms or services units, modular stair/lift cores, and modules for disaster relief. Timber-framed modules have been used in hotels, schools, and housing, but are less widely used than steel or concrete modules.

4.1 TIMBER-FRAMED MODULES

Timber-framed modules of various forms are used in 1- and 2-storey educational buildings and in housing, which are described below.

4.1.1 Educational buildings

Specialist companies produce a range of timber modules for temporary and permanent classrooms, sports halls, dining halls and kitchens, day nurseries, laboratories, and medical and office buildings. Typical specifications for timber-framed modules in this sector are:

- Floors: 100×50 mm timber floor joists with 18 mm plywood or particleboard glued and nailed to the timber joists.
- External walls: 100×50 mm timber frame clad with Stoneflex cladding and with 60 mm closed-cell insulation inserted between the timber studs. The internal lining is 12.5 mm plasterboard.
- Roof: Timber box beams at 2.4 m centres spanned by timber joists laid to a longitudinal slope and covered by a roof deck of plywood or similar board that is finished in water-roofed felt.

Terrapin's timber Unitrex system differs from other modular systems in that it is delivered to the site "flat packed"; floor panels are installed first, and roof panels added to support timber posts. Wall panels are then inserted. Roofs can be flat or pitched, and a one-hinged pitched solution is also available, as shown in Figure 4.1.

4.1.2 Housing

Timber modules for housing consist of 38×89 mm timber studs with 9 mm oriented strand board (OSB) sheathing board fixed on the outside face. The walls are insulated externally with rigid insulation board and are covered with a vapour-permeable waterproof membrane. Mineral wool insulation is placed between the timber studs, and 12.5 mm impact-resistant gypsum board forms the internal surface.

A brick façade is separated from the timber structure by a 50 mm wide cavity and closed-cell insulation boards are placed outside the modules. Details are shown in Figure 4.2. The manufacture of the timber floor is shown in Figure 4.3. Its depth is typically 385 mm when using 250 mm deep floor joists.

An example of the installation of timber modules in a 2-storey semi-detached house is shown in Figure 4.4. In this case, two 4 m wide modules form one house, and the modules are manufactured with internal stairs, kitchens, bathrooms, and partitions. The roof is formed using conventional timber trusses. The completed modular housing project with its conventional brick cladding and tiled roofing is shown in Figure 4.5.

4.2 REUSE OF SHIPPING CONTAINERS

There is a surplus of shipping containers in Europe due to the imbalance between imports and exports to the Far East, as it is not economical to ship empty containers. They are designed to transport goods of all kinds by ship and lorry and are made from steel frames comprising hollow corner sections and welded corrugated steel walls. Standardised lifting points are situated at the corners of the containers (see Figure 4.6).

The dimensions and structural properties of shipping containers mean that they can easily be converted to many temporary or permanent uses, stores, etc. Their primary advantage is that they are readily available and can be transported by conventional container lorry without escort. Some companies specialise in custom

Figure 4.1 Flat-pack modular timber panels used at Stantonbury School, Milton Keynes. (Courtesy of Terrapin.)

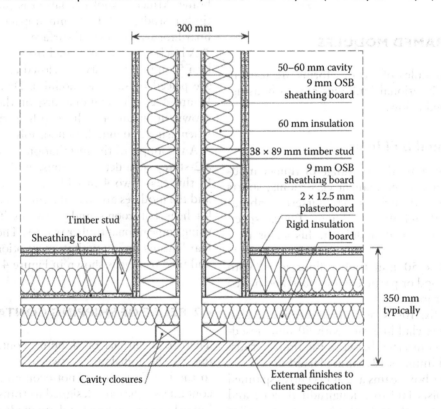

Figure 4.2 Details of timber module. (Courtesy of Hunter Offsite Ltd.)

conversions of redundant shipping containers, and examples of office and display spaces in London using containers are shown in Figures 4.7 and 4.8.

Containers are a standard external width of 2.42 m (8 ft), and range from 2.42 m (8 ft) to 12.19 m (40 ft) in length and have heights of 2.59 m (8.5 ft) and 2.89 m (9.5 ft) in external dimensions. Taller containers can be obtained, but their external width is always 2.42 m.

More ambitious use has been made of containers to form whole or parts of multistorey buildings. A sports hall was constructed in just 3 days for Dunraven School in Streatham, London, using shipping containers to form three sides of the building. The fourth side was fully glazed. The 3-storey-high containers also support the long-span steel roof trusses. The containers house changing rooms, viewing galleries, and toilets. One important feature of this design was that potentially the whole building could be moved depending on the future requirements of the school. The external and internal views of the completed building are shown in Figure 4.9.

Figure 4.3 Timber floor plate being manufactured. (Courtesy of Hunter Offsite Ltd.)

Figure 4.6 Lifting of shipping containers from their corners. (Courtesy of Matthias Hamm.)

Figure 4.4 Timber modules being assembled on site. (Courtesy of Hunter Offsite Ltd.)

Figure 4.7 Containers used for workshops and offices, London Docklands.

Figure 4.5 Completed timber-framed modular housing. (Courtesy of Hunter Offsite Ltd.)

Freitag Individual Recycled Freeway Shop is a 26 m high concept store in Zurich that was built using 17 shipping containers. A modular staircase also using containers forms the tower. Large windows are provided at the ends of the containers together with a fully a glazed entrance area. Although designed as temporary, the shop has been in place longer than its original 10-year expected life.

An early example of the use of container units in a residential building was in a 3-storey student residence in Delft, Netherlands, shown in Figure 4.10.

A theatre in Cologne, Germany, was built using portable and temporary modules for the back of the theatre changing rooms and offices. A side view is shown in Figure 4.11. The theatre was covered by tubular steel arches and a flexible membrane roof, so that in theory,

Figure 4.8 Group of refurbished containers used for a restaurant, London Waterloo.

Figure 4.9 New sports hall and gymnasium using containers to support the roof. (Designed by SCABEL Architects.)

the whole building could be reconfigured or dismantled and moved in the future. This is one of the underlying benefits of modular systems.

Another good example of the use of containers was in the Verbus system, which has been used for hotels, such as a 9-storey hotel in Uxbridge, west London, shown completed in Figure 4.12.

4.3 BATHROOM PODS

Pods are non-load-bearing modular units that are usually highly serviced. They are often manufactured as bathrooms and kitchens or combinations of both, and are made in a variety of materials, such as glass reinforced plastic (GRP), polyester, thermoplastics, precast concrete, steel panels, or light steel framing faced with marine-grade composite boards, fibre-reinforced cement building boards, or similar. A typical layout of a bathroom pod is shown in Figure 4.13, which is typically 2 m × 2.4 m in external dimensions. These types of pods are normally produced by specialist manufacturers in a range of standard designs, but who can also tailor-make their bathroom pods for relatively large production runs. An example of the high quality of finishes that are possible is shown in Figure 4.14.

Concrete pods are structurally rigid but are relatively heavy. Light steel-framed pods provide a lighter-weight solution and are sufficiently rigid for their size and use. Light steel bathroom modules can be used as structural elements in combination with light steel framing, as shown in Figure 4.15. Elements Europe has developed

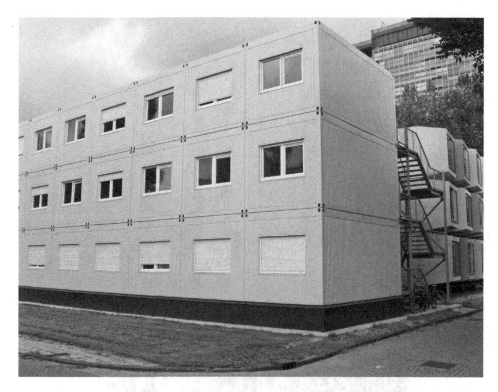

Figure 4.10 Student residential building in Delft, the Netherlands, built using containers.

Figure 4.11 Musical Dome theatre in Cologne using modular units for the back of the theatre.

Figure 4.12 Hotel in Uxbridge, West London, constructed using shipping containers. (Courtesy of Verbus.)

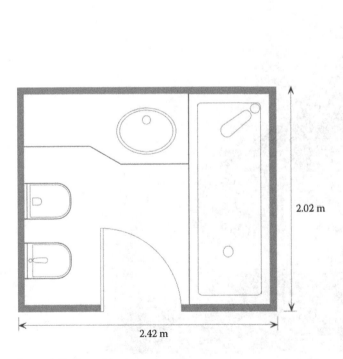

2.02 m

2.42 m

Figure 4.13 Typical bathroom pod layout.

Figure 4.14 Internal view of modular bathroom. (Courtesy of Caledonian Modular.)

Figure 4.15 Light steel load-bearing bathroom modules.

Figure 4.17 PVC bathroom pod positioned on a light steel floor. (Courtesy of Offsite Solutions and Metek.)

Figure 4.16 Installation of modular bathroom unit. (Courtesy of Elements Europe.)

its Strucpod system that uses a bathroom module as part of the load-bearing structure. A bathroom module being lifted into place is shown in Figure 4.16.

GRP pods are very lightweight and are watertight, but have less structural rigidity. An example is shown in Figure 4.17. Floorless pods minimise the depth of the construction to avoid the problem of "stepping into" a bathroom module.

Specialist pods for toilets and kitchens are often installed within conventional concrete or steel-framed structures in educational and healthcare buildings, commercial buildings, and hotels. Large bathroom/toilet units have been used in city centre commercial buildings since the mid-1980s.

The pods are craned to the required level onto scaffold landing platforms or onto floor-level gantries that are cantilevered out from the floor structure. Hoist platforms can also be used to lift the pods from ground level. The pods can then be moved from the landing area to their planned locations by either using pallet trucks or, in some models, using special wheel assemblies supplied by the pod manufacturer. Shims or pads are used to adjust the level to align with the finished floor level. The main building services and drainage system are connected to the connection points on the exterior wall of the pods.

4.4 SPECIAL FORMS OF MODULAR CONSTRUCTION

Modules are often designed for specialist applications, such as

- Lifts and stairs
- Balconies
- Exhibition spaces
- Rooftop extensions
- Micro-living space

Stair modules can be manufactured with a partially open top and base, as illustrated in Figure 4.18. In this case, the upper flight of stairs projects above the ceiling level so that it is level with the floor of the module above. Alternatively, stair modules can be designed as a rigid welded frame often using square or rectangular hollow sections, as shown in Figure 4.19. This module was used in the rooftop extension of an existing building shown on page 58. Details of stair modules are described in Chapter 14.

Figure 4.18 Stair module with corner posts. (Courtesy of Kingspan Steel Building Solutions.)

Figure 4.19 Stair module using welded RHS members being lifted into place. (Courtesy of Powerwall.)

Figure 4.20 Modules used for exhibition space. (Courtesy of Caledonian Modular.)

Figure 4.21 Pull-to-open module for disaster relief. (Courtesy of CSM, Rome, Italy.)

Specialist modules can be manufactured for exhibition buildings, as in Figure 4.20, which is a steel and glass module manufactured for a housing developer in London. The design eludes to the Farnsworth house by Mise van der Rohe. It consists of a floor and roof cassettes made from 300 mm deep light steel C sections that are supported by 200 mm deep double C sections as columns located in-board of the corners of the module. Three modules formed a one open-plan single-storey exhibition space.

The pull-to-open module was developed in Italy for use in disaster relief. The module can be transported on local roads in its 2 m width and then extended sideways by a cog and rail system to create a 3.6 m wide room with a 1.8 m wide internal corridor space and fitments along the extended sides. The module is 6 m long and has extendable legs for use on variable ground levels. It is shown in its demonstration form in Figure 4.21.

Microcompact home (m-ch) is a concept developed by TU Munich, and shown in the case studies, which is a 2.6 m cube designed for single-person living. Another system is loft cube, which is a modular system that is intended to be lifted onto flat roofs.

4.5 MODULAR UNITS IN RENOVATION

Modular units are often used in renovation projects to provide new space by extending buildings horizontally or vertically. Modular bathrooms may be stacked up to 10 storeys high from ground level and are supported laterally by the existing building. New balconies may be attached between the modules. An example of a load-bearing bathroom unit used in renovation is shown in

Figure 4.22 Modular bathrooms used in building renovation. (Courtesy of Ruukki.)

Figure 4.23 Modular rooftop units with a glazed façade used in building renovation in Copenhagen and also showing a module being lifted into place.

Figure 4.22. The bathroom modules can be finished in a lightweight cladding designed to match the colour of the existing concrete façade.

Modules may be placed on the roof or on structural walls of an existing building and are used to create a rooftop extension, as shown in Figures 4.23 and 4.24. The rooftop modules may also comprise a sloping roof. In this way, the time for completion of the rooftop extension is dramatically reduced from many months to a few days.

Figure 4.24 Installation of rooftop modules and completed renovation of a 1960s housing block in west London. (Courtesy of Powerwall.)

Figure 4.25 Four-storey modular lift used in building renovation in Helsinki. (Courtesy of NEAPO.)

Figure 4.26 Corefast lift module as delivered to site and installed. (Courtesy of Tata Steel.)

New external stairs in modular form may also be provided to access the new rooftop floor. The use of external modular lifts is illustrated in Figure 4.25.

4.6 ACCESS CORES

Corefast is a special form of steel module that can be manufactured in both 2D and 3D form to create cores in multistorey buildings. The double-skin steel panel can be prefabricated and installed as 1- or 2-storey high elements, as shown in Figure 4.26(a). It is later filled with concrete to provide composite action under wind loads and for fire resistance. A completed modular core is shown in Figure 4.26(b). Precast concrete cores are described in Chapter 3.

CASE STUDY 11: CONTAINERS FOR HOTEL, UXBRIDGE, WEST LONDON

View of completed hotel, Uxbridge.

The first hotel in Europe to be constructed from modified steel shipping containers opened in the centre of Uxbridge, west London, in August 2008. The 9-storey Travelodge hotel was built from 86 standard steel shipping containers that were used to create 120 bedrooms. The container modules are 2.68 m wide and 3 m high and 12 m long externally, which included the corridor. The containers were delivered as part of the Verbus system. They were adapted and fitted out with the hotel's standard fixtures and furnishings in Shenzen, China, and were shipped to the UK.

Without using modular construction, it is unlikely to have been possible to construct a hotel on the site, which is next to the Uxbridge main bus station. The placing of the 86 container modules took only 20 days, ensuring minimum disruption to the local area. Two different sizes of container modules were used to build the hotel, creating double rooms that measure 5 × 3 m and also family rooms that measure 3.5 × 6 m. Disabled rooms were also available.

The container modules are made of high-strength corrosion-resistant steel, and the walls, floors, and roof construction were insulated for sound and thermal performance. The modules are designed to be stacked in several configurations. The sizes available allow designers to connect or nest modules and provide the stability of the resulting structure. The potential for multistorey uses of container modules is currently 16 storeys.

The resulting structure provides support for external parts and communal space, such as cladding, roof systems, stairwells, corridors, balconies, and access areas. A system of standard connections allows interconnecting components to interface with foundations, cladding systems, roofing, balconies, and corridor units.

For medium-sized hotels—those with more than 200 rooms—Verbus claims its modules are up to 20% cheaper and 50% faster than traditional building systems. The internal partitioning of the container to create a two-room hotel module is shown above. Hotels in Heathrow and Warminster have also been constructed using this system.

CASE STUDY 12: MODULAR UNITS IN ROOFTOP EXTENSION

Completed building with its rooftop extension.

Existing concrete panel building.

The rooftop extension of existing buildings is an important market for modular construction. A good example of the renovation of a series of 1960s precast concrete buildings is located on Du Cane Road in Shepherds Bush, west London. The site is hemmed in between the main road and the Piccadilly underground line, and is opposite the Queen Mary Hospital, which meant that there was no space for storage or off-loading from the road.

The project consisted of the renovation of five existing 3-storey Bison concrete panel buildings. Each had a central stair core and deck access to the two wings of each building. The client, the Du Cane Housing Association, wished to minimise disturbance to the residents, and ensured that one block at a time was made available for renovation. Contractor Apollo Property Services Group chose modular construction because it minimised the impact of the building process, and reduced labour-intensive site work, materials handling, and storage.

The new floor of each of the buildings comprised 20 rooftop modules in one- and two-bedroom configurations. The modules were typically 9 m long and 3.5 m wide and were orientated along the axis of the buildings. One module formed the front of the new floor, and one formed the rear, so that the windows and doors were inbuilt into one side of each module. A prefabricated steel balcony was attached to the modules on site before being lifted into place.

The modules comprised 70 mm deep C sections for the walls and 150 mm deep sections for the floor joists and roof. They were designed to provide high levels of thermal and acoustic insulation to satisfy the Code for Sustainable Homes level 4. Two modules formed a single-bedroom apartment of 56 m^2 floor area to meet Lifetime Homes standards. A pair of longer modules formed a two-bedroom apartment of 65 m^2 floor area. Two infill blocks were built using load-bearing modules and comprised three bedroom units in double-module forms.

Modules were installed at a rate of eight per day, and the rooftop modules in one block were installed over 2 days in successive weeks. The infill blocks consisting of 44 modules were constructed in only 3 weeks. Each module weighed approximately 8 tonnes. At this relatively light load, it was calculated that the additional load on existing structure and foundations was less than 10% of the existing building, and therefore the new floor could be supported without strengthening the precast concrete structure.

The overall value of the project was £10.5 million, which included comprehensive refurbishment of the 112 apartments in the existing five blocks to modern standards and energy performance. A total of 44 new apartments were created comprising 150 modules, making the project the largest known use of modular construction in renovation or building extensions.

CASE STUDY 13: MICROCOMPACT HOME, AUSTRIA AND GERMANY

Microcompact home in Munich.

Tree concept using microcompact homes.

The microcompact home (m-ch) is a lightweight compact dwelling for one or two people. It was developed by architect Richard Hordern. Its compact dimensions of a 2.66 m cube allow it to be adapted to a variety of sites and circumstances, and its sleeping, working/dining, cooking, and sanitation spaces make it suitable for everyday use. Informed by aviation and automotive design and manufactured at the production centre in Austria, the m-ch can be delivered with project individual graphics and interior finishes.

The m-ch has a timber-framed structure with anodised aluminium external cladding, insulated with polyurethane and fitted with aluminium frame double-glazed windows and front door with security lock. It is 2.66 m high externally, with an internal ceiling height of 1.98 m. M-ch modules can be stacked vertically and clustered together around a common external stair access. A m-ch unit weighs 2.2 tonnes and its internal features include

- Two compact double beds
- A sliding table measuring 1.05 × 0.65 m, for dining
- A shower and toilet cubicle
- A kitchen area

An open-core space contains the central lift shaft and stairway surrounded by 30 microcompact homes. These are supplied with power and water from an internal ring of vertical services "reeds." The microcompact homes are arranged around the core in a way to provide maximum transparency and openness for nature to penetrate the space.

M-ch units are available to purchase for delivery in Europe and the United States at a guide price from €38,000, which includes an aluminium subframe, stair and balustrades, and interior fittings. This price is also subject to site conditions and excludes delivery, installation, foundations, connection to services, consultant's fee, and taxes. The m-ch delivery time is an average of 8 to 10 weeks from order.

A village of seven microcompact homes, sponsored by international telecoms company O2 Germany, was built in 2005 at the Technical University Munich. A 15 m high tree village is also planned, based on a 12 m² footprint to fit a mature landscape with tall trees. Its structure is made up of a cluster of small steel vertical columns or reeds.

Chapter 5

Introduction to planning of modular buildings

Design of buildings using modular construction is a complex inter-relationship between the desired space and function of the building and the economical use of similar-sized modules. Modular construction requires a new discipline based on the use of large building blocks rather than use of skeletal or planar components with which designers are familiar. An optimised modular system should allow for flexibility in internal planning, but must retain the discipline of off-site manufacture in terms of standardisation of components and manufacturing efficiency.

5.1 GENERAL PRINCIPLES

Certain general principles apply when designing buildings using modular units, which also extend to the parts of the building that are constructed using other forms of construction. This may include the floor panels for the corridors and circulation areas, the stairs, main services, cladding, and roof. These principles are

- Decide whether four-sided modules satisfy the spatial and functional requirements, or whether open-sided modules are required to achieve more effective space use.
- Design the building layout to achieve as much repetition as possible in the size and fit-out of the modules. The load-bearing capacity of the modular structure can be varied while maintaining the same external geometry.
- Choose the module size to be compatible with transport, local access, and installation constraints. For transportation, the maximum module width is typically 4.2 m, but the module length can be up to 16 m (see Chapter 17).
- Decide how the building may be stabilised by using the group of modules alone, or in combination with additional bracing, or for high-rise buildings, by a concrete or braced steel core.
- Prefit the services and equipment within the modules and decide how these services are accessed from the outside of the modules, and how they are distributed through the building.

- Consider the fire safety strategy and effective fire compartmentation provided by a group of modules. Modules with two layers of plasterboard achieve 90 min of fire resistance.
- Consider the cladding system to be used and how it may be connected to the modules. Decide whether the joints between the modules are to be emphasised or hidden as part of the architectural concept.

The plan forms that may be considered at the concept design stage of modular buildings fall into well-defined types, which are described as follows.

5.2 CORRIDOR-TYPE BUILDINGS

The most commonly used modular construction system comprises a linear assembly of four-sided modules placed on either side of a central corridor, as illustrated in Figure 5.1(a). This arrangement is ideally suited for hotels, student accommodation, etc. From a structural point of view, wind forces on the front and rear façades are resisted efficiently by the side walls of the modules, whereas wind forces on the end gables are resisted by the highly perforated and hence weaker façade walls. This means that for efficient design, the building should be longer than it is deep. A simple relationship between the number of modules in a group and the building height is shown in Table 5.1.

The maximum height of corridor-type modular buildings depends on the particular system and wind exposure, but 5 to 7 storeys is considered to be the limit for modular buildings without additional bracing. An additional stabilising system may be introduced in the form of concrete cores or steel bracing around stairs, as illustrated in Figure 5.1(b). In this case, buildings up to 25 storeys high are feasible, depending on the plan form and the particular modular system used. It is often necessary to strengthen the structure of the lower modules, where loads are highest, while retaining the same external geometry of the modules on all levels.

Figure 5.2 shows a 12-storey corridor-type modular building, which provides for mixed use as a student residence with commercial space below. The 6 to

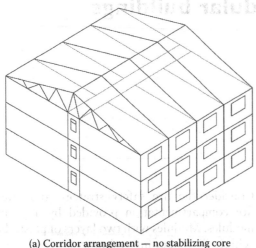

(a) Corridor arrangement — no stabilizing core

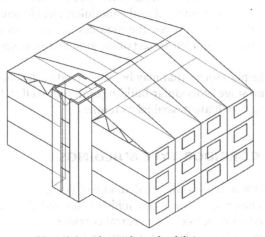

(b) Module with corridor and stabilizing core

Figure 5.1 Corridor-type building form using groups of modules.

Table 5.1 Minimum number of modules required on the front elevation for corridor-type buildings

Building height (number of storeys)	Minimum number of modules along front façade	Separate stabilising system required
$N_s = 3$	5	No
$N_s = 4$	7	No
$N_s = 5$	9	No
$N_s = 6$	11	Possibly
$N_s = 7$	12	Possibly
$N_s = 8$	12	Yes

10 storeys of modules are supported on a 2-storey steel-framed podium structure. The 400 standard bedroom modules are 2.7 m wide externally, but approximately 100 modules are combined in pairs to form premium studios consisting of two rooms. The kitchen modules are 3.6 m wide externally.

For student residences, the group of five modular bedrooms and a communal kitchen module are arranged with a double corridor, which provides an acoustic and a fire separation function, so that the group of six modules effectively comprises one apartment. In this building, stability was provided by four braced steel cores, into which some modules are placed (see the plan form in Figure 5.3).

The supporting steel structure at the first floor level is designed so that the beams align with the load-bearing walls of the modules. The columns are typically arranged on a 6 to 8 m grid to support pairs of modules above, which leads to effective use of the office space below. This type of construction is described in more detail in Chapter 10.

5.3 EXTERNAL ACCESS BUILDINGS

5.3.1 Two- and three-storey housing

Two- and three-storey townhouses can be designed using groups of two, three, or four modules with access at the ground level, and with either independent stairs (normally contained within a separate stair and lift module) or integral stairs contained within the room modules. These techniques are explored for housing in Chapter 6. An example of a terrace of 3-storey town houses built using 10 m long modules is illustrated in Figure 5.4.

5.3.2 Multistorey modular buildings

In multistorey buildings with external access, a separate steel structure provides the external walkways to the modules, and this structure may be located on one or both sides of the line of modules. This form of construction was first used in the Murray Grove project shown in Figure 1.1. The external walkway is designed to resist the vertical loads and wind forces applied to it.

A single stack of modules is limited in height to 5 or 6 storeys, unless another stabilisation system is used, such as a braced access core. In its simplest form, an external access structure is supported by the modules and by tubular columns, with additional bracing or lateral support by the stair core, as shown in Figure 5.5. In more elaborate systems, the external steel structure can be used to stabilise a group of modules.

Modules may also be arranged longitudinally, as was the case in the Moho project in Manchester, which represented a breakthrough in residential building design. It was designed with partially open-sided modules, and the external steel walkways and balconies were designed as an "exoskeleton" to resist wind forces applied to the whole building. This arrangement is illustrated during construction in Figure 5.6, in which the modules were

Figure 5.2 Twelve-storey modular student residence at Bond Street, Bristol. (Courtesy of Unite Modular Solutions.)

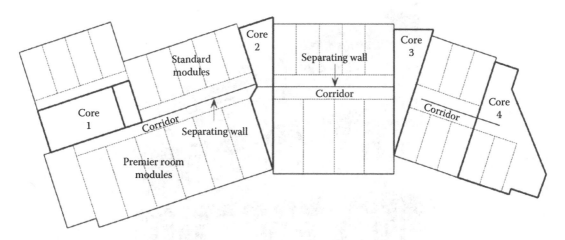

Figure 5.3 Plan of module building at Bond Street, Bristol, showing the double-corridor layout and the core positions.

placed after the steelwork had been erected. The completed building is shown in Figure 1.3.

An adaptation of this technique may be used to create an atrium-type building with gallery access between the modules. The atrium may be an independent structure or may be supported by the modules, as shown in the 3-storey waiting area and gallery circulation space at the Hull Royal Infirmary in Figure 5.7. In this case, the walkways are made as part of the modules and the corner posts are designed to support the additional walkway and atrium roof loads.

5.4 OPEN-PLAN MODULAR BUILDINGS

Open-plan space may be created by using modules with corner posts and relatively deep edge beams. The maximum longitudinal span of an open-sided module is typically 12 m, although larger modules are available from some manufacturers. This type of open-sided module is often used for schools, commercial, retail, and medical buildings. The walls of modules with corner posts are non-load bearing unless they contribute to the stability

Figure 5.4 Town houses in Twickenham using modular construction. (Courtesy of Futureform.)

Figure 5.5 Rear view of a group of modules with an external access structure supported by intermediate posts. (Courtesy of Caledonian Modular.)

Figure 5.6 Open-sided module with corner and intermediate posts supported by a structural frame. (Courtesy of Yorkon and Joule Engineers.)

Figure 5.7 Open-sided module with corner posts used to create a walkway and atrium structure at Hull Royal Infirmary. (Courtesy of Portakabin.)

Figure 5.8 Open-sided module used in a hospital building. (Courtesy of Yorkon.)

under lateral loads. Intermediate posts may be introduced to minimise the depth of the edge beam. The corner posts of a group of modules form a larger column when used in open-plan space.

This form of construction may be used with any arrangement of modules, but is limited in height unless cores or braced walls are placed strategically at key locations on the building plan. Horizontal forces are transferred through the connections between the modules and diaphragm action of the floors and ceiling to the vertical bracing or core. Modules can be reorientated at the corner posts if the module length is manufactured as a multiple of its width, as illustrated in Figure 5.8.

Schools require open-plan space for classrooms. For example, three 3 m wide by 9 m long open-sided modules create a 9 × 9 m classroom. In the Yorkon system, three 3.75 m wide by 16.5 m long open-sided modules can be used efficiently to create two classrooms, an integral corridor/cloakroom, and a store room. In some systems, an intermediate post can be located at the line of the integral corridor to reduce the span of the edge beam in an open-sided module.

5.5 HIGH-RISE MODULAR BUILDINGS

Modular construction is conventionally used for buildings with multiple similar rooms of up to 10 storeys height where the walls are load bearing and also resist shear forces due to wind actions. More recently, modular construction has been extended up to 25 storeys height by using additional concrete cores or structural frames to resist wind loading and to provide stability

for the group of modules. Therefore, the modules are designed to resist the accumulated compression loads over the building height, but not the horizontal forces.

One technique is to cluster modules around a concrete core, which houses the stairs and lifts, so that the wind forces applied to the modules are transferred horizontally to the core walls. This concept has been used on major residential projects, such as Paragon in west London, shown in Figure 5.9, which was the first high-rise use of modular construction (Cartz and Crosby, 2007). In this project, the modules were constructed with corner posts, and at the lower floors, the sizes of the corner posts were increased by using square hollow sections (SHSs) of the same external width but increasing thickness. Other high-rise modular buildings have been completed in north London and Wolverhampton, and are illustrated in Figures 1.6 and 1.18.

The concrete core may be constructed by slip forming or jump forming at the same time as the modules are placed. This speeds up the construction process, although the modular units are installed vertically at a much faster rate than the core. Steel plates are cast into the core so that the modules may be later connected (generally by welding) to these plates in order to transfer the required lateral forces.

The building form may be elongated laterally provided that wind loads can be transferred to the concrete core. This can be achieved by using in-plane trusses placed within the corridors, or by consideration of the structural ties between the modules and their attachment to the core. Various alternative high-rise building forms in which modules are clustered around a core are presented in Chapter 6.

Figure 5.9 Modular building stabilised by a concrete core in construction and completed. (Courtesy of Caledonian Modular.)

5.6 DIMENSIONS FOR PLANNING OF MODULAR BUILDINGS

The factors that influence the dimensional planning of modular systems in building design may be summarised as

- Building form, as influenced by its requirements for access, circulation, and communal space
- Planning grid for internal fitments, such as kitchen units
- Transportation requirements, including access and installation (see earlier)
- Alignment with external dimensions of cladding, e.g., brick dimensions
- Efficient utilisation of space, which influences the floor and wall widths

The planning grid depends on the use of the building, and the following internal planning dimensions are widely used:

- Offices: 1500 mm
- Hospitals/schools: 1200 mm
- Housing: 600 mm

A dimensional unit of 300 mm is generally adopted as the standard for vertical and horizontal dimensions, reducing to 150 mm as a second level for vertical dimensions.

Some modular suppliers have developed their own planning grid for efficient use of their modules when grouped together for different applications. For example, the Yorkon system is based on a 3.75 × 7.5 to 18.75 m configuration, which allows modules to be reorientated on a 3.75 m planning grid.

The geometric standards that may be used for concept design of modular buildings are based broadly on the following dimensions:

- Wall width of 300 mm for internal separating walls and external walls
- Floor depth of 450 mm for the combined floor and ceiling depth in modular and hybrid construction systems
- Floor depth of 600 to 750 mm for corner-supported modules
- Internal planning dimensions based on 600 mm (therefore 3 or 3.6 m is the preferred internal modular width)
- Floor-ceiling heights based on 2.4 m for residential buildings and 2.7 or 3 m for commercial, health, or educational buildings

In residential applications, a floor to ceiling height of 2400 mm aligns with plasterboard sizes. Internal wall heights will be 300 or 600 mm greater for schools, health centres, and commercial buildings.

The width and length of the rooms tend to be governed more by the use of the space. The following *internal* module widths are used for common applications:

- Study bedrooms: 2.5 to 2.7 m
- Schools (open-sided modules): 3 m
- Hotels, social housing: 3.3 to 3.6 m
- Apartments, offices: 3.6 m
- Health sector buildings: 3.6 to 4 m

However, actual widths may vary, depending on efficient space utilisation of the building on the particular site, and so the above internal dimensions are only indicative. Externally, modules will be 250 to 300 mm wider than their internal dimensions.

Transportation requirements may also influence the final decision on module size.

The module lengths can be designed in larger increments, such as 600 mm, and the following lengths are typically used for the following applications:

- Study bedrooms and hotels: 5.4 to 6 m
- Apartments, social housing: 7.5 to 9 m
- Primary schools: 8.4 m
- Secondary schools: 9 to 12 m
- Offices: 6 to 12 m
- Supermarkets, health sector buildings: 10 to 12 m

Module sizes have increased in response to customer demand for larger modules, particularly in the educational, health, and retail sectors. The module length is generally not as important for transportation as the width, except where site access is difficult. However, the longer the span of the edge members in open-sided modules, the greater is the overall floor depth.

5.7 STRUCTURAL ZONES

5.7.1 Internal walls

The width of the pair of internal walls in modular systems may be designed for scheme purposes as 300 mm, which incorporates the various boards and insulation (see Figure 5.10(a)). The gap between the walls is a variable, depending on the number and thickness of boards and size of the wall studs. Most modular systems lead to wall dimensions in the range of 250 to 300 mm, and so the planning dimension of 300 mm can be achieved in practice.

5.7.2 External walls

External walls are dimensioned according to the type of cladding. A total external wall width of 300 mm may be adopted as a guide for most types of cladding materials, depending on the amount of insulation used externally to the module. The actual wall width will vary, from 230 mm for insulated render to 300 mm for rain screen cladding systems to 380 mm for brickwork.

Brickwork design is based on a standard unit of 225 mm length and 75 mm depth. Therefore, in modular buildings, it is important to design the overall floor-to-floor depth as a multiple of 75 mm in order to avoid non-standard coursing of bricks. A floor-to-floor height in multiples of 150 or 300 mm clearly achieves this requirement. A multiple of 225 mm in horizontal brickwork coursing width is more difficult to achieve, both in the module width and in the window size.

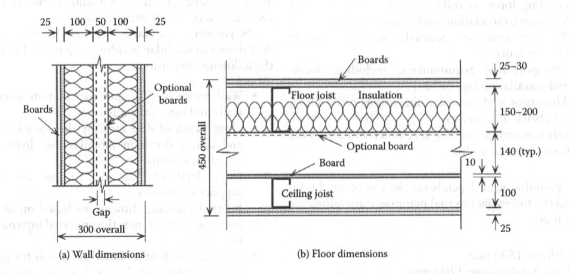

(a) Wall dimensions   (b) Floor dimensions

Figure 5.10 Wall and floor dimensions for planning in modular construction.

An external module width of 3600 mm (allowing for gaps between the modules) fits with this brickwork coursing in a plain façade, but problems still exist at corners and brickwork returns, etc. It is possible to overcome these dimensional constraints by using a variable-width cavity at the corners of the building, but this will also require the use of longer brick ties at these positions.

Other types of cladding, such as clay tiles or metallic panels, have their own dimensional requirements, but generally they can be designed and manufactured to fit around windows and at corners, etc. Many types of lightweight cladding can be preattached to the modules, but if so, it is generally necessary to install a cover piece over the joint between the modules on site to allow for geometrical tolerances.

5.7.3 Floor zone

Floors and ceilings in modular construction are deeper than in more traditional construction. The three structural cases noted earlier will require different overall ceiling-floor dimensions for planning purposes, as follows:

- Continuously supported modules: 300 to 450 mm
- Corner-supported modules: 600 to 750 mm
- Frame-supported modules: 750 to 900 mm

In most cases, 450 mm may be adopted as a standard for the combined floor and ceiling dimension, as in Figure 5.10(b) (Lawson, 2007). However, many modular systems achieve shallower depths, and a combined floor and ceiling zone of 300 or 375 mm is feasible in some systems that align with brick coursing.

The details of a corner-supported module are illustrated in Figure 5.11. In this case, a combined floor and ceiling depth of 600 mm may be used in planning, depending on the depth of the edge members. The gap between the floor and ceiling is a variable depending on

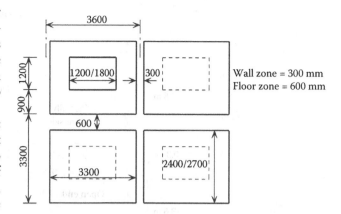

Figure 5.12 Summary of standardised module dimensions for planning purposes in building design.

the floor and ceiling joist sizes, and the minimum gap is usually 20 mm.

Typical planning dimensions for layout of modules are summarised in Figure 5.12. Actual external dimensions of the modules will be less than these planning dimensions to allow for gaps between the modules. Windows and doors may also be incorporated as standard conventional dimensions.

5.8 OPEN BUILDING APPROACH

An open building approach, as defined by CIB Working Group 104, has the aim of achieving flexibility in internal planning and servicing both in the initial design and in future changes of use. While in modular construction a high degree of design flexibility has to be balanced against the manufacturing requirements, it is possible to provide a building system based on floor grid positions in which module orientations and sizes can be varied to provide flexible use of the space.

The use of modular construction may also be optimised by manufacturing the higher-value parts of the building as modular units, and the open-plan space is built using planar or skeletal elements. In this way, modules are used for the serviced spaces, such as kitchens and bathrooms. This is explored further in Chapter 10.

The layout of modules to create an open system using internal posts based on a 3.6, 3.75, or 3.9 m grid is illustrated in Figure 5.13. The posts may be in the form of hot-rolled steel angles or square hollow sections (SHSs) (typically 100 × 100 mm) that are combined to form an internal column. The modules may be manufactured with partial open sides to optimise the use of the internal space.

The plan view shows the possible orientation of the modules, assuming that intermediate posts are provided so that the free span of the edge member in a partial open-sided module is 3.6 m. For this span, the

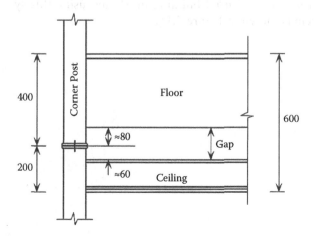

Figure 5.11 Typical dimensions of a corner-supported module.

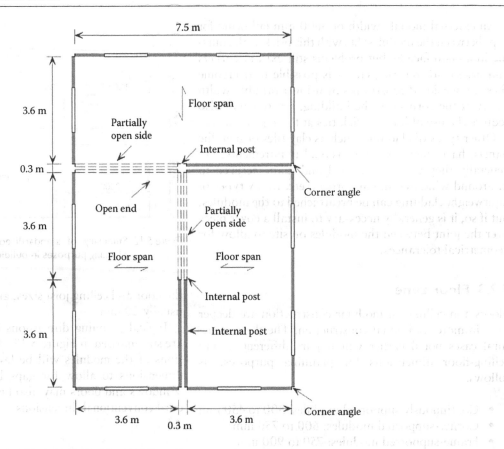

Figure 5.13 Creation of flexible space using partially open-sided modules.

edge beam depth is the same as the floor. Balconies can be attached to the SHS corner or side posts. Non-load-bearing walls can be positioned anywhere on the plan.

In this concept, the module length should ideally be a multiple of two or three times its width (therefore, 7.5 or 11.25 m for a 3.75 m grid) in order to facilitate the reorientation of the modules, as shown in Figure 5.13. The cluster of posts at changes of orientation of the modules can be "boxed out" on site. SHS posts are recommended for buildings more than 4 storeys high, as they are more stable in compression when unsupported by walls in one or both directions.

A separate stabilising system in the form of braced bays is generally required, and X-bracing can be installed between the posts in walls, without modifying the basic open building system.

This open building approach was applied in a Swedish system, Open House (Lessing, 2004), where modules are supported by 100 × 100 mm SHS posts placed on a 3.9 m grid. The posts are external to and not part of the modules, as illustrated in Figure 5.14. The corners of the modules were recessed to incorporate the SHS posts. Using the 3.9 m grid, modules could be reorientated at post positions. Balconies, stairs, and other features may be attached to the posts. A completed 4-storey residential building in Malmo using this system is shown in Figure 5.15.

Figure 5.14 Completed modular housing in Malmo, Sweden. (Courtesy of Open House AB.)

Figure 5.15 Installation of modules with recessed corners around
SHS posts. (Courtesy of Open House AB.)

CASE STUDY 14: HIGH-RISE APARTMENTS AND STUDENT RESIDENCE, LONDON

Completed 11-storey student residence in west London.
(Courtesy of Caledonian Modular.)

Installation of double-length module at the lower levels.

Developer Berkeley First chose modular construction for its key worker and starter homes project called Paragon in Brentford, west London, because of its short construction programme of 22 months and to minimise logistical problems on site. The £26 million project provided one- and two-bedroom accommodation for key workers and for students at Thames Valley University. The 17-storey building was completed in September 2006. Sandwiched between the M4, local housing, and a school, the site presented difficulties for access, delivery, and storage of materials, which modular construction solved.

The extension of the use of modular construction up to 17 storeys in this project was achieved by use of a concrete core, which provided overall stability of the buildings. In this way, the modules are designed to resist only vertical loading through their corner posts, and wind loads are transferred to the core. Modules were attached to the concrete core by steel angles fixed to channels cast into the concrete. Construction of the slip-formed cores was completed in advance of the modules being installed. In some areas, the modules were installed on a steel podium in order to allow vehicular access below to the basement level.

The project comprises six separate buildings of 4, 5, 7, 12, and 17 storeys. Caledonian Modular manufactured a wide range of module types, many with open sides and integral corridors. The total number of modules in the project was 827, and the 17-storey building consists of 413 modules. A total of 600 en-suite student rooms, 114 en-suite studio rooms, and 44 one-bedroom and 63 two-bedroom key worker apartments were provided. A typical module size is 12 × 2.8 m, but some were manufactured up to 4.2 m wide.

The modules used 80 × 80 SHS or 160 × 80 RHS sections in varying thicknesses at the corners of the modules, depending on the building height. These posts fit within the light steel walls. The combined floor and ceiling depth was 400 mm, and the combined width of walls was 290 mm. Both achieved airborne sound reduction of over 60 dB and a fire resistance of up to 120 min.

The edge beams use 200 × 90 parallel flange channels (PFCs) at the floor level and 140 × 70 PFCs at the ceiling level in order to design partially open-sided modules of up to 6 m span. The one- or two-bedroom apartments were constructed using two or three modules, each of 35 to 55 m² floor area. The long modules included the corridor, which speeded up the construction process.

CASE STUDY 15: PRIVATE APARTMENTS WITH INTEGRAL BALCONIES, MANCHESTER

External view of the balconies and enclosed space of the MOHO building.

Installation of modules with three large patio windows. (Courtesy of Yorkon.)

A 7-storey apartment building for developer Urban Splash, called MoHo (short for modular housing), was the first modular building in the UK built purely for sale rather than social housing for rent. The U-shaped building is located in the Castlefield area of Manchester. It consists of six residential floors built over one commercial and retail floor and two levels of basement car parking. The 102 apartments are configured in one-bedroom and two-bedroom formats, and each has its own enclosed balcony space, internal bathroom, and kitchen.

Architect ShedKM, modular specialist Yorkon, and structural engineer Joule devised a novel design in which modules are arranged parallel rather than perpendicular to the façade and are constructed with an open side using intermediate posts. The balconies and access walkways were constructed first as a conventional steel frame and provided the stability to the building, and all horizontal loads are transferred through the modular connections to the external steel structure. In this way, the modules could be designed with partial or fully open sides, which enhances their "airy" feel. The living space also extends into the enclosed balconies.

The apartment sizes varied from 38 to 54 m² excluding balconies. The apartments were laid out in various formats by using modules oriented parallel to the building façades. The external dimensions of the modules were 4.1 × 9.1 m. Modules were manufactured with one or two intermediate posts in the form of 100 mm square hollow sections. Two modules created the two-bedroom units, and all were supplied with fully glazed side walls.

The largest apartment of 12.1 m length was achieved by extending the basic module using an additional second bedroom module. Kitchens and bathrooms were manufactured as internal pods or islands within the modules. The room space is increased by use of a dining pod extending into the enclosed balcony and an internal entrance lobby pod.

Loads are transferred through the corner and the intermediate square posts of the modules to the transfer structure. The external structure of the balconies and access walkways was constructed with careful location of the X-bracing, and the frame was connected to the corners of the modules in order to transfer loads between them.

The Yorkon modules took only 5 weeks to install at a rate of six per day having erected the steel-framed exoskeleton. Modules were delivered "just in time" to suit local traffic conditions. The as-built construction cost, including basement car parking, was £1330/m² floor area. The construction period for the whole project was 17 months, which saved an estimated 7 months.

CASE STUDY 16: MODULAR APARTMENTS AND HOME OFFICES, DUBLIN

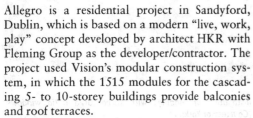

Installation of modules on a concrete podium.

Internal view of living room.

Allegro is a residential project in Sandyford, Dublin, which is based on a modern "live, work, play" concept developed by architect HKR with Fleming Group as the developer/contractor. The project used Vision's modular construction system, in which the 1515 modules for the cascading 5- to 10-storey buildings provide balconies and roof terraces.

Block A consists of 224 apartments in one-, two-, and three-bedroom configurations. The one-bedroom apartments use two modules, the two-bedroom apartments use three modules, and the three-bedroom apartments use four modules. Shared corridors were also of modular manufacture. The home office apartments at the ground floor level use four or five modules, two of which are for office use.

The irregular plan form was achieved using modules offset along their length and even manufacture of non-rectangular-shaped modules. A number of the modules were designed with partially open sides so that larger rooms and corridor spaces could be created. A total of 730 modules in block A were installed in just 20 weeks, having completed the concrete cores.

The buildings are supported by a reinforced concrete podium above two basement car parking levels. The external cladding is in the form of granite panels fixed to the outward façade walls of the horseshoe-shaped buildings. The modules support the imposed loads and the external cladding, but the overall stability is provided by concrete lift/stair cores.

The Vision modular system consists of concrete floors supported by parallel flange channel (PFC) sections. The walls use 60 × 60 square hollow sections (SHSs) placed at 600 mm centres, which support the 9 storeys of modules. Partially open-sided modules utilise the inherent spanning capabilities of the PFC edge beams in the floor. The floors and walls also achieved the 120 min fire rating, which was necessary for this project.

The ground floor modules provide office space, and all apartments have private balcony space built as a separate steel framework but supported by the modules. A three-bedroom configuration used four modules, accessed directly from the communal staircase. The module widths vary from 3.3 to 4.2 m and lengths from 6 to 11 m. A main bathroom and en suite bathroom was provided in all two- and three-bedroom apartment formats. The internal height of the modules was generally 2.45 m, increasing to 3 m in the ground floor modules.

Modules arrived on site almost fully finished internally with most services in place. Connections to adjacent modules and from the modules into the stair or lift cores were completed on site. Kitchens and bathrooms were completed with all fixed units in place.

CASE STUDY 17: OPEN HOUSE SYSTEM, MALMO, SWEDEN

Open House modular construction in Malmo, southern Sweden. (Courtesy of Open House AB.)

Installation of L-shaped module. (Courtesy of Open House AB.)

The Open House concept is a multistorey residential system, developed in Sweden, that gives considerable architectural freedom. These features have been utilised in several projects, as demonstrated by the two projects in Helsingborg and Malmö, southern Sweden. Annestad was a very large development and a total of 1200 apartments were built from 2004 to 2008. The development was divided into medium-sized 2½- to 5-storey blocks. The development was a combination of rental apartments and tenant-owned apartments. The rental cost is about €110 per m²/year, and the initial market price is about €1500 to 1800 per m² floor area.

The modules were arranged around, and were supported by, six steel columns on a 3.9 m grid. Four square hollow section (SHS) columns were located in the corners, and two or three along the sides of the modules. The internal measurements of the modules are 3.6 m by 7.2 m up to 11 m (i.e., 300 mm is allowed for the combined wall widths). The modules can cantilever 1.7 m from the exterior columns, without changing the basic 3.9 m grid arrangement. The typical weight of a fitted-out module is 5 to 8 tonnes.

The size of the apartments varies from one room to four rooms plus a kitchen and bathroom. Modules are positioned in an offset configuration to create a variable façade line. Façade materials used in this project were combinations of bricks, boards, insulated render, wood, and steel panels.

The modules were fully serviced and were delivered from the factory near Malmo. Many modules were manufactured with open sides. Some L-shaped modules were manufactured, as shown above. For most of the buildings, the façades, roofs, and balconies were constructed on site.

The modules use slotted steel C sections in the walls, complemented with mineral wool and exterior and interior plasterboards. The exterior walls provide a high level of thermal performance by a U-value of 0.15 W/m²K. Light steel sections, mineral wool, gypsum board, and trapezoidal steel sheets are used for the roof and floor of the modules. Good sound insulation and fire resistance are provided.

The installation procedure first required the installation of the SHS posts on pad foundations on a 3.9 m grid, and the modules with recessed corners are placed between the columns and connected to them. This system meant that the modules could be manufactured with partially or fully open sides, while maintaining a minimum depth of floor.

REFERENCES

Cartz, J.P., and Crosby, M. (2007). Building high-rise modular homes. *Structural Engineer*, 85(19).

CIB (International Council for Research and Innovation in Building Construction). *W104: Open building implementation*. www.cibworld.nl.

Lawson, R.M. (2007). *Building design using modular construction*. Steel Construction Institute P348.

Lessing, J. (2004). *Industrial production of apartments with steel frames. A study of the OpenHouse system*. Report 229-4. Swedish Institute of Steel Construction, Stockholm, Sweden.

Chapter 6

Housing and residential buildings

The design of housing and multistorey residential buildings in modular construction is strongly dependent on the room sizes that are required, their spatial arrangement, and the opportunities to use similar-sized modular units. Modular construction is well established in the hotel, student residence, and military accommodation sectors, in which the room sizes are essentially the same throughout the project. Variants are often only left- and right-handed configurations. In the housing and residential sector, more design flexibility is required while maintaining the discipline of the off-site manufacturing process.

This chapter reviews the spatial and other design requirements for housing and residential buildings, the means of satisfying the Building Regulations in the UK, and the opportunity to use modular units in various plan forms. Applications in high-rise residential buildings are presented.

6.1 SPACE PLANNING IN HOUSING

The most widely used standards for space planning in housing and residential buildings in the UK are

1. Design and Quality Standards (Housing Corporation, 2007), which replaced the Scheme Development Standards (2003)
2. Lifetime Homes (Joseph Rowntree Foundation, 2010)
3. Standards and Quality in Development (National Housing Federation, 2008)

The National Housing Federation (NHF) standards give minimum room sizes for different occupancy patterns, and other space, energy efficiency, and performance requirements. Other requirements that are relevant are given in the "London Housing Design Guide—Interim" (Mayor of London, 2010) and in the Housing Quality Indicators for schemes funded by the Homes and Communities Agency (www.homesandcommunities.co.uk/hqi).

These standards and also Lifetime Homes provide layouts compatible with the new requirements for disabled access in Building Regulations (Approved Document M). Typical dwelling sizes and the likely number of modules that are required to comply with the NHF and the London Housing Guide layouts are in the range of

1 bedroom, 1 person	30–35 m²	1 module
1 bedroom, 2 person	45–50 m²	2 modules
2 bedroom, 3 person	57–67 m²	3 modules
2 bedroom, 4 person	67–75 m²	3 modules
3 bedroom, 5 person	75–85 m²	4 modules

Typical room sizes are summarised in Table 6.1 for various occupancy levels.

In modular construction, room sizes are dependent on the sizes of the modular units, which should be within the geometric limits for transportation. Typically two modules form a two-person apartment of approximately 50 m² floor area. Because of the nature of modular construction, bedrooms and living rooms are usually of the same width (typically 3 to 3.6 m internally). In some cases, the modules are manufactured with integral corridors and balconies and with partially open sides so that wider living spaces can be designed.

The design of residential developments for a secure environment is addressed in an initiative called Secured by Design. In practice, it means that doors and windows have to meet minimum standards of security and the overall development has to make good use of lighting and surveillance.

6.2 BUILDING REGULATIONS (ENGLAND AND WALES)

Guidance on meeting the Building Regulations in England and Wales is given in a series of approved documents. Recent regulatory changes have introduced stricter requirements for thermal insulation and for acoustic insulation of separating floors and walls.

Table 6.1 Typical space requirements for dwellings

Room	Space requirements (m² floor area) for number of people per dwelling				
	1	2	3	4	5
Living room	11	12	13	14	15
Living/dining	13	13	15	16	17.5
Kitchen	5.5	5.5	5.5	7	7
Kitchen/dining	8	9	11	11	12
Main bedroom	8	11	11	11	11
Twin bedroom	—	—	10	10	10
Single bedroom	—	—	6.5	6.5	6.5

Source: Joseph Rowntree Foundation, Lifetime Homes, 2010.

Similar requirements exist in Scotland, which has its own regulations.

6.2.1 Thermal insulation

The thermal insulation requirements for housing and residential buildings are given in Building Regulations (England and Wales) Approved Document L1 (2010). To comply with the regulations, the dwelling CO_2 emission rate (DER) should not exceed a target CO_2 emission rate (TER). The DER is calculated using the energy required for heating, lighting, etc., less any savings due to use of renewable energy systems.

The heat transmittance is characterised by the U-value of a unit area of the external envelope, and its units are W/m² per °C temperature difference across it. The maximum permitted U-values were reduced in the 2010 Building Regulations, and will reduce further in 2013. The target U-values that are required to achieve the TER are presented in Table 6.2. These U-values should not be confused with the backstop U-values given in the regulations, which are the higher and are maximum permitted values for any given element of the building envelope.

In all types of building construction, it is necessary to ensure that "cold bridging" does not cause disproportionate heat losses or risk of condensation. Cold

Table 6.2 Progression of target U-values of elements of building envelope of housing in Building Regulations (2006–2013)

Element	U-value (W/m²°C)		
	2006 Regulations	2010 Regulations	2013 Regulations (planned)
External walls	0.30 (gas heating)	0.25 (gas heating)	0.20 (gas heating)
	0.25 (electric heating)	0.20 (electric heating)	0.15 (electric heating)
Ground floors	0.22	0.20	0.20
Roofs (pitched)	0.16	0.15	0.15

bridging can occur if structural elements penetrate the building envelope, and in this case, more detailed calculations of local heat loss are required.

Airtightness is also an important parameter, as studies have shown that unwanted air infiltration can increase heat losses significantly. An airtightness test in the UK is carried out to a relatively high-pressure differential of 50 Pa, and a test permeability of 10 m³/m²/h is considered to be representative of normal building practice. In practice, actual air infiltration through the building fabric in normal conditions will only be about 5% of the test value.

Modular units are more airtight than similar on-site construction, which is partly due to the use of sheathing boards and sealed joints between the internal and external boards. If required, membranes can be introduced into the manufacture of the modules. Typically modules achieve an airtightness of 2 to 3 m³/m²/h.

To achieve a U-value of 0.2 W/m²°C in both light steel and timber modular construction, it is necessary to introduce 100 mm of inter-stud insulation (often in the form of mineral wool) as well as up to 80 mm of closed-cell insulation board placed outside the framework (see Chapter 14 for guidance on insulation levels for modern cladding systems).

6.2.2 Future regulations

A report by the Zero Carbon Hub (ZCH), *Defining a Fabric Energy Efficiency Standard for New Homes* (2007), presented a range of solutions to achieve a 25 to 30% reduction in heating energy use in buildings compared to the 2006 Building Regulations. This will be the target for the 2013 Building Regulations, and corresponds to a maximum space heating requirement of 39 W/m² floor area per year for mid-terrace houses and apartments, and 46 W/m²/year for semi- or detached or end of terrace houses, which reflects their higher exposed surface areas.

The optimised specification for the building fabric was based on the minimum capital cost less the net present value of the energy savings over 60 years, based on real increase in energy costs of 2.5% per year plus inflation. The target specification for the building fabric that was proposed by the ZCH for new buildings to achieve the space heating requirements is presented in Table 6.3. Also shown is the equivalent Passive House planning requirements. The thermal bridging parameter, y, takes account of the accumulated heat losses at all thermal bridges and is added to the heat loss in the building fabric (i.e., external walls, roof, and ground floor).

The proposed air permeability of 3 m³/m²/h through the building envelope is much better than in current regulations and can be achieved in modular construction. In houses of this level of airtightness, mechanical

Table 6.3 Thermal characteristics required to achieve 25% to 30% energy reduction relative to current practice (Zero Carbon Hub)

Thermal parameter	House types		
	All house types except detached	Detached houses	Passive house standard
External walls, U-value	0.18 W/m²K	0.18 W/m²K	0.10–0.15 W/m²K
Ground floor, U-value	0.18 W/m²K	0.14 W/m²K	0.10 W/m²K
Roof, U-value	0.13 W/m²K	0.11 W/m²K	0.10 W/m²K
Windows, U-value	1.4 W/m²K	1.3 W/m²K	0.8 W/m²K
Doors, U-value	1.2 W/m²K	1.2 W/m²K	0.8 W/m²K
Thermal bridging parameter, y	0.05 W/m²K	0.04 W/m²K	0.04 W/m²K
Airtightness	3 m³/m²/h	3 m³/m²/h	0.5 m³/m²/h

Source: Zero Carbon Hub, *Defining a Fabric Energy Efficiency Standard for New Homes*, 2009, http://www.zerocarbonhub.org.

ventilation and heat recovery (MVHR) systems are installed to maintain air quality and reduce energy losses. MVHR units are normally incorporated in the roof space, but can also be built in to the modular units.

6.2.3 Fire safety

Modular residential buildings often have relatively high occupation density, and their cellular nature means that the fire safety of the modules has to be considered individually and as a group.

Safety in the event of fire is achieved by measures to allow the occupants to escape safely and to ensure effective firefighting. This is achieved in practice by compartmentation to prevent fire spread, by clear and alternative means of escape, by use of incombustible materials, and by suitable fire resistance that is dependent on the building height and function. The regulatory requirement for the use of sprinklers in residential buildings is limited to buildings whose top habitable level is greater than 30 m above ground, and in some hotels and mixed-use buildings. The design of residential buildings is strongly influenced by the layout of apartments and the travel distances to stairs or fire-protected lobbies.

The regulations for fire safety are embodied in Approved Document B, and BS 9999: *Code of Practice for Fire Safety in the Design, Management and Use of Buildings*, which presents general requirements for all types of buildings. The specific requirements for residential buildings are given in BS 9991: *Fire in the Design*

Managements, and Use of Residential Buildings: Code of Practice, which replaces BS 5588-1: *Fire Precautions in the Design, Construction and Use of Buildings: Code of Practice for Residential Buildings*, which gave more specific guidance for residential buildings. Effective means of escape is achieved by one of the two main approaches in residential buildings with a corridor and fire-protected lobby:

1. By limiting the travel distance from the exit door of an apartment to a smoke-free area
2. By provision of alternative means of escape to a smoke-free area

If a staircase is used for firefighting or is in a residential building, a vent is required in the protected area for smoke control and the staircase generally should also have a vent.

Cross-corridor self-closing fire doors should be provided in long corridors that can be in the form of hold-open devices for accessibility that are triggered by an alarm/detection system. Other effective measures that may be considered are pressurisation of escape routes (to prevent smoke access) and use of sprinklers (to reduce fire spread).

The travel distance is limited to a maximum of 7.5 m from the exit of the dwelling to the entrance to a fire-protected stairway or lobby. If this is not satisfied, separate fire-protected doors are required in the corridor to satisfy the maximum distances from the exit of the dwelling to the protected area. Active measures of smoke control would generally be required if these distances are exceeded. Maximum travel distances are presented in Figure 6.1 for two floor layouts, one being a corridor access and the other being access from a lobby. Some relaxations are permitted for small buildings less than 11 m high (i.e., 4 storeys).

Fire resistance is required to ensure that the structure remains stable in fire, depending on the height of the building, which is defined as to the top of the highest floor. Current fire resistance requirements are defined in Table 6.4, together with the approximate number of storeys for each fire resistance class. For mixed-use buildings, other fire resistance and means of escape requirements may apply.

Modular buildings are generally designed with two 15 mm thick layers of plasterboard internally to satisfy the acoustic insulation requirements for separating floors and walls. The same build-up of boards generally achieves 90 min fire resistance, which is required for buildings up to 10 storeys high. External sheathing boards and fire stops in the cavity also assist in reducing the passage of smoke in fire. For taller buildings, a further layer of plasterboard is required to achieve 120 min of fire resistance, and this also improves the acoustic insulation between the modules.

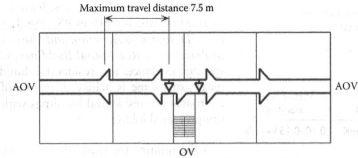

(a) Maximum travel distances for corridor access to lobby

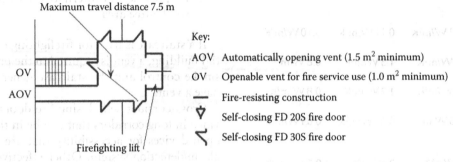

Key:

AOV   Automatically opening vent (1.5 m^2 minimum)

OV   Openable vent for fire service use (1.0 m^2 minimum)

—   Fire-resisting construction

∇   Self-closing FD 20S fire door

⟨   Self-closing FD 30S fire door

(b) Maximum travel distances for common access to lobby

Note 1. Where all dwellings on a storey have independent alternative means of escape the maximum distance may be increased to 30 m.

Note 2. Where a firefighting lift is required, it should be sited not more than 7.5 m

Note 3. The OVs to the stairway may be replaced by an openable vent over the stair.

Figure 6.1 Minimum escape distances in residential buildings.

Table 6.4 Fire resistance requirements in Approved Document B

	Fire resistance (min)			
Parameter	R30	R60	R90	R120
Maximum height (m)[a]	<5 m	<18 m	<30 m	>30 m
Maximum number of storeys[b]	2	6	10	>10

[a] Defined to top of highest floor.
[b] Typical depending on floor zone.

6.2.4 Acoustic insulation

The acoustic performance requirements for buildings for residential purposes are given in Approved Document E. Acoustic requirements and performance data are presented in Chapter 11. In general, the double-layer walls and the combined floor and ceiling in modular construction perform well acoustically.

6.3 HOUSE FORMS IN MODULAR CONSTRUCTION

The design of houses and residential buildings in modular construction depends on satisfying the functional use of the space within the dimensional requirements of modular manufacture. Some examples of modern modular housing are shown in Figure 6.2.

Modules can be manufactured with partial open sides to facilitate the more flexible use of space, as illustrated in Figure 6.3. Modules can also be manufactured with integral balconies, service risers, etc. Furthermore, the use of preinstalled services in the modules means that access for service connection between the modules and for future maintenance has to be considered. The following sections describe the various building forms that can be designed using modular construction.

The simplest assembly of modules in a simple family home consists of two modules per floor in a 2-storey house, as illustrated in Figure 6.4(a). Two of the modules provide the open living space and bedrooms, and two provide the kitchen, bathroom, and stairs.

The typical internal width of a room module in housing is 3.3 m, corresponding to an external width of 3.6 m. For housing, the efficient use of space often requires that the kitchen/bathroom/stairs modules are narrower (typically 3 m width) than the adjacent room modules. The house frontage is therefore approximately 6.6 m. The module length is variable depending on the house layout, and is typically 8 to 10 m.

A 3-storey house may be created by manufacturing the roof modules as a mansard shape, as shown in Figure 6.4(b). A 3-storey terrace of town houses using

(a)   (b)

Figure 6.2 Housing using modular construction. (Courtesy of Futureform.) (a) Housing in Harlow, Essex, UK. (b) CUB house at the Building Research Establishment, UK.

Figure 6.3 Internal view of modular building in Figure 6.2(a).

six modules per house is illustrated in Figure 6.5. In this case, the modules are 3.6 m wide and 10 m long. The brickwork cladding is ground supported and is tied to the modules using brick ties that are attached to vertical runners connected to the modules.

It is also possible to design a terraced house of narrow frontage using a single wide module per floor. A possible layout of a modular house using two modules is illustrated in Figure 6.6. The modules are 4.2 m wide and 10 to 12 m long (external dimensions) and have integral stairs. A third floor can be added without changing the basic modular system. In the manufacture of the modules, the vertical service route from the bathroom above should be aligned with kitchen. Sufficient access should be provided to connect the services vertically and for future maintenance. This dictates the position of the services zone on plan, which in the case shown is on the bathroom side of the internal wall.

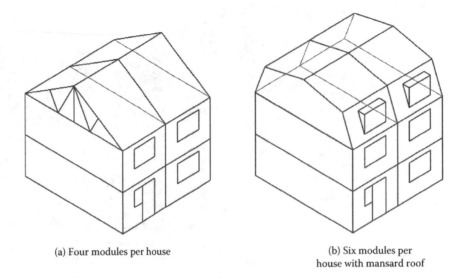

(a) Four modules per house

(b) Six modules per house with mansard roof

Figure 6.4 Simple forms of housing using four modules.

Figure 6.5 Three-storey town houses in east London. (Courtesy of Rollalong and Metek.)

The CUB Housing system developed by Futureform consists of two modules per floor, but in this case, the modules are placed transverse to the house frontage. Panels in the floor and ceiling allow for stairs to be inserted, and additional modules can be added at a later date in a novel generation house concept. The room layout and structural system are shown in Figure 6.7, which is extendable up to 4 storeys. The completed CUB house at the BRE Innovation Park is shown in Figure 6.2(b).

6.4 RESIDENTIAL BUILDINGS IN MODULAR CONSTRUCTION

Various arrangements of modules may be considered in the efficient layout of apartments, depending on whether they are

- Accessed externally, i.e., deck access
- Corridor-type buildings
- Modules clustered around a concrete core

Figure 6.6 Plan form of 2-storey house using single 4.2 m wide modules per floor.

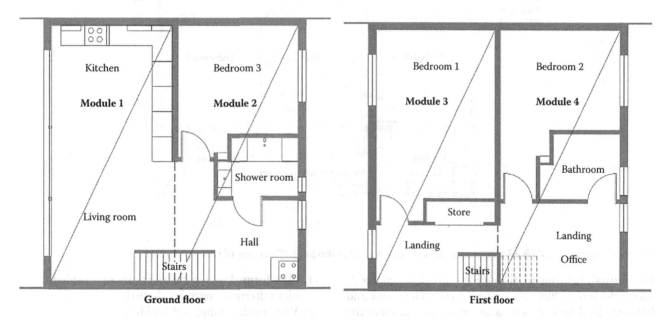

Figure 6.7 CUB Housing using modular construction showing the ground floor and upper floor layout. (Courtesy of Futureform.)

A single apartment usually comprises two modules for a single-bedroom configuration and three modules for a double-bedroom configuration, each with a single bathroom and possibly a further en suite bathroom in three-module configurations. Modules can be manufactured with or without integral balconies as part of the modules. Various configurations of apartments with integral balconies are illustrated in Figures 6.8 and 6.9. In all cases, the internal module width is 3.3 m, which is acceptable to Lifetime Homes standards. Service risers are accessed from the outside of the module.

The simplest arrangement of modules is in a corridor-type layout, and an example of this format comprising four apartments per floor is illustrated in Figure 6.10. In this layout, two relatively large modules each form a two-bedroom apartment. In corridor-type layouts, the degree of fenestration has to be sufficiently high

Figure 6.8 Apartment layout consisting of two 3.6 × 7.2 m modules. (Courtesy of HTA Architects.)

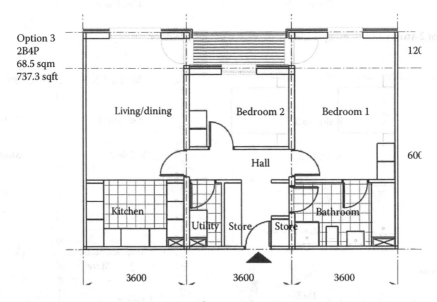

Figure 6.9 Apartment of 69 m² area consisting of three 3.6 × 7.2 m modules. (Courtesy of HTA Architects.)

for natural lighting of the living room/kitchen modules, which only have one external end. The stairs and lifts may be designed as separate modules that serve the four apartments.

Partially open-sided modules lead to more flexible use of the internal space, as illustrated in Figure 6.11. In this social housing project in south London, the stairs, and potentially also a lift, are incorporated as a separate module, which serves two apartments per floor. The apartments consist of two modules of different lengths to create single- and two-bedroom variants. Figure 6.11 shows the

1. Plan form showing the partially open-sided modules (alternate modules shaded)
2. View of the completed building

An efficient arrangement of nine ground floor modules was achieved in the Birchway project in west London, shown in Figure 6.12. This innovative project consists of five similar 2-storey buildings and was the first modular project to achieve the Code for Sustainable Homes level 5 sustainability rating (see case studies). The ground floor plan is shown in Figure 6.13, in which the hall is also manufactured as a module. The upper

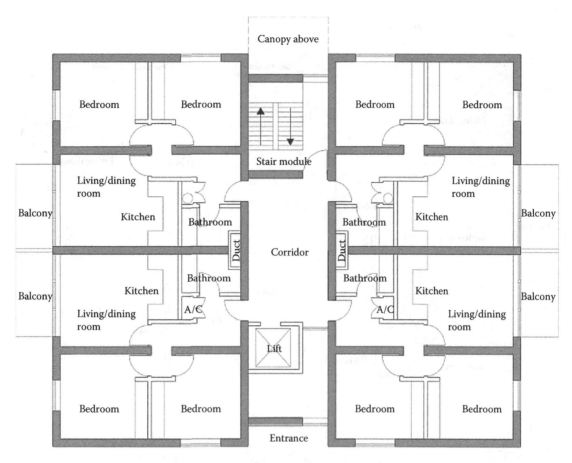

Figure 6.10 Example of corridor layout of apartments in modular construction. (Courtesy of Caledonian Modular.)

floor comprises four modules, which support a curved light steel roof structure and a "green" sedum roof. Photovoltaic panels were placed on the south-facing side of the roof.

More complex building forms can be created by careful location of walls in a regular module form, as was done in the Paragon project in west London (see Figure 6.14). In this case, some of the partially open-sided modules were manufactured with square hollow section (SHS) posts set back from the corners so that the corridor could be manufactured as part of the modules. The completed building is shown in Figure 6.15.

6.5 STUDENT RESIDENCES

Modules for student residences are generally constructed in light steel framing, but some are constructed in precast concrete (see Chapter 3). In student residences, the modules are normally arranged in a corridor style, as illustrated in Figure 6.16. Study bedroom modules are often relatively small, and are typically 2.7 m wide and 6 m long externally.

A group of five study bedrooms and a single communal kitchen is considered to be one unit for acoustic separation and fire compartmentation purposes. This often leads to the use of a double-separated corridor layout. Longer or wider modules can be designed as studio rooms. Despite the relatively simple plan form of a student residential building, there are seven different module types in the plan form of Figure 6.16, taking account of left- and right-handed variants. Longer or wider modules can be designed, as shown in Figure 6.17.

Examples of multistorey student residences are illustrated in Figures 6.18 and 6.19. Often a ground floor podium is constructed in in situ concrete or as a steel framework to provide communal space at the ground floor. In Figure 6.19, the colour scheme of the second floor plan indicates the different room sizes and specifications. The communal kitchens are generally 3.6 m wide, and larger studio rooms were included in this project, which is shown completed in Figure 6.20. The grey areas in this plan form show the access cores. The pairs of corridors allow for independent access to a group of modules.

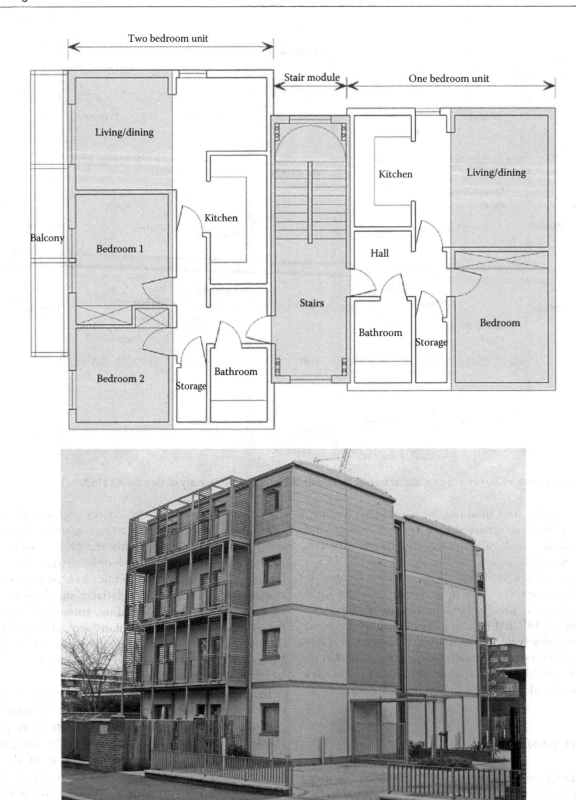

Figure 6.11 Modules create flexible use of space in Stockwell, south London. (a) Plan form showing partially open sides (alternate modules are shaded). (b) View of completed building. (Courtesy of PCKO Architects.)

Figure 6.12 Completed building at Birchway, west London, showing its curved green roof. (Courtesy of Futureform.)

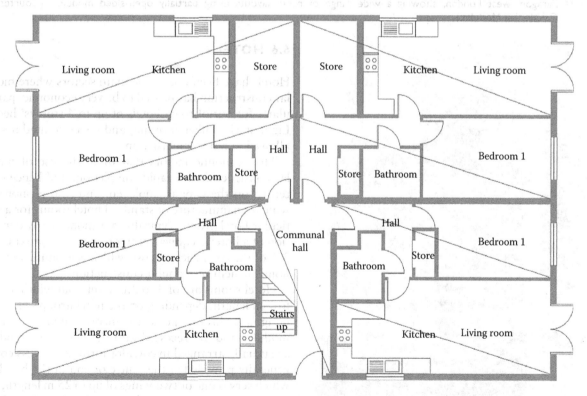

Figure 6.13 Layout of the four apartments on the ground floor of the building in Figure 6.12 using modules of different sizes. (Courtesy of Futureform.)

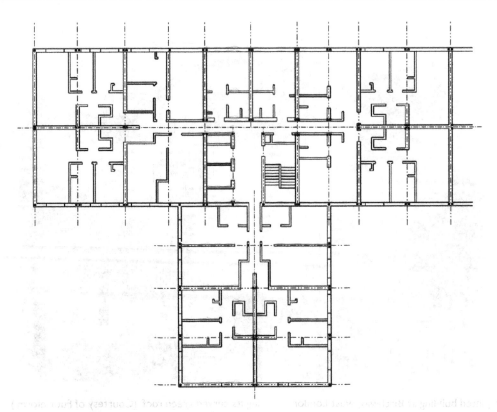

Figure 6.14 Paragon, west London, showing a wide range of room layouts using partially open-sided modules. (Courtesy of Caledonian Modular.)

Figure 6.15 Façade of Paragon in west London. (Courtesy of Caledonian Modular.)

6.6 HOTELS

Hotels have been one of the main sectors where modular construction has proved to be very economic, particularly for out-of-town hotels of up to 4 storeys' height. Light steel, timber framing, and concrete modules are all used in hotel construction.

The economic factors that drive the use of modular construction are rapid completion, and hence early return on the capital employed, and also economy of scale in manufacture of standard hotel rooms for a particular hotel group. Typically, a 3-month early completion of a hotel is equivalent in revenue to approximately 3% of the construction cost, which is a significant economic benefit of modular construction.

Hotel rooms are of 3 to 3.6 m internal width and up to 7 m length, depending on the hotel group's standard design. Bathroom pods are incorporated within the modules and are recessed into the floor. The modules are usually arranged in corridor form. An access core is generally provided at the centre or end of the building, which serves one or two wings of up to 25 m length, and comprising up to 16 modules per wing. Separate means of escape is often required at the ends of the corridors.

A recent trend has been the construction of inner city hotels in which 4 to 6 storeys of modular rooms are supported on a steel frame at the first floor level so that the reception and restaurant are located in the open-plan

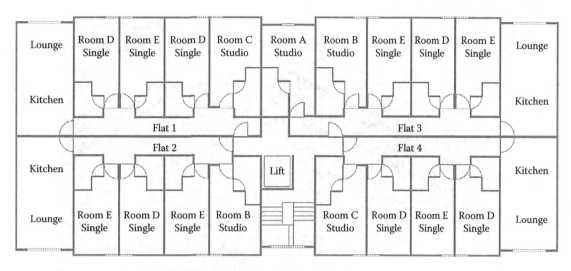

Figure 6.16 Typical room layouts in the form of flats for compartmentation purposes in a student residence.

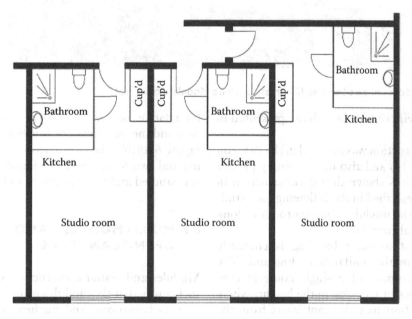

Figure 6.17 Typical studio room layouts with a variant of a longer module.

space on the ground floor. A good example in central London is illustrated in Figure 6.21, in which the columns are aligned with the walls of alternate modules.

6.7 MODULAR LAYOUTS IN TALL BUILDINGS

For medium- to high-rise buildings, it is often more efficient to cluster the modules around a concrete core, which provides the overall stability of the building. In this building form, the transfer of forces from the modules to the core is minimised. A configuration of modules using this principle is illustrated in Figure 6.22,

in which eight apartments comprising 16 modules are placed around the core and are accessed from it. The concrete core is usually constructed first, usually as an in situ slip-formed concrete, or sometimes using precast concrete core units (see Chapter 4).

Taller buildings can be extended horizontally from the core by using horizontal bracing placed in the corridors. In this case, the forces transferred via the bracing to the core can be relatively high. Steel plates can be cast into the core during its construction, and a further steel plate may be welded to it in order to form the connection to the corridor and to the modules. Individual modules remote from the core are connected to the cor-

Figure 6.18 Student residence in east London. (Courtesy of Unite Modular Solutions.)

ridor structure at their corners. Details are presented in Chapter 12.

This form of construction was adopted in the Paragon project (see Figure 6.14) and also in a 16-storey project near Wembley, which is shown during construction in Figure 6.23 and is described in the following case studies. In this project, the modules were up to 16 m long and included a central corridor.

The world's tallest modular building is currently in Wolverhampton in the midlands of England. The 25-storey building consists of a single concrete core and five distinct zones of modules in the height, where a group of three or four modules cantilevers from the steel-framed structure below. It is shown during construction in Figure 6.24(a). The plate attachments to the top of the modules are welded on site, and the connections to the core are placed in vertical slots to allow

for relative movement over the long term between the core and the modules. The finished building is shown in Figure 6.24(b). The construction process for this building and two 8- and 9-storey buildings on the same site was studied and is summarised in Chapter 19.

6.8 MIXED MODULES AND OPEN-PLAN SPACE

Modules and planar construction may be combined in such a way that the modules provide the serviced zones, such as bathrooms and kitchen, and the planar floor elements provide the open-plan spaces. This leads to efficient space planning and reduces much of the dimensional constraints of modules. This is discussed in more detail in Chapter 10.

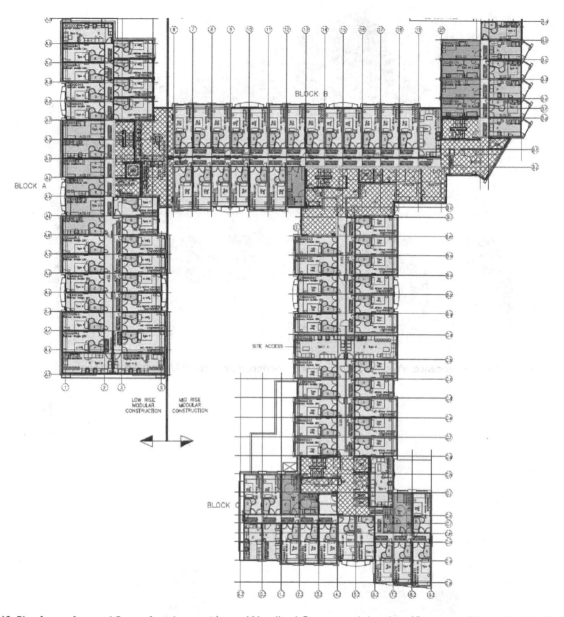

Figure 6.19 Plan form of second floor of student residence, Woodland Court, north London. (Courtesy of Unite Modular Solutions.)

Figure 6.20 Modular student residence, Woodland Court, north London. (Courtesy of Unite Modular Solutions.)

Figure 6.21 Installation of modules on a steel podium for a hotel in central London. (Courtesy of Futureform and Citizen M hotels.)

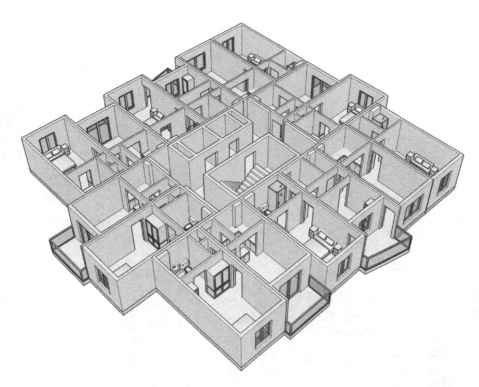

Figure 6.22 Typical room layout for cluster-type building in high-rise buildings. (Courtesy of HTA Architects.)

Figure 6.23 High-rise modular building in Wembley using an in situ concrete core. (Courtesy of Futureform.)

(a)   (b)

Figure 6.24 Twenty-five-storey modular building in Wolverhampton (a) during construction and (b) completed. (Courtesy of Vision.)

CASE STUDY 18: MODULAR HOUSING, HARLOW, ESSEX

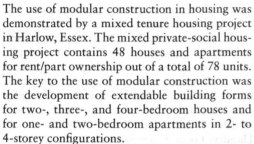

View of private housing. (Courtesy of Futureform.)

Internal view of living area.

The use of modular construction in housing was demonstrated by a mixed tenure housing project in Harlow, Essex. The mixed private-social housing project contains 48 houses and apartments for rent/part ownership out of a total of 78 units. The key to the use of modular construction was the development of extendable building forms for two-, three-, and four-bedroom houses and for one- and two-bedroom apartments in 2- to 4-storey configurations.

A total of 177 house modules and 72 apartment modules were installed at a maximum rate of 10 per day. The modules, designed by Futureform, use the Ayrframe fabrication system by Ayrshire Metal Products and were fully fitted out in Futureform's assembly plant in Wellingborough, Northants. The finishing work on site was limited to foundations, cladding, roofing, and service connections.

Sister company Renascent Developments Ltd. was one of the joint development partners that formed the development company South Chase New Hall Ltd., and MOAT Housing Association has acquired the affordable housing elements. These organisations took an active part in the procurement and construction process in order to maximise the benefits of off-site manufacture. Architect Proctor and Matthews has also worked closely with Futureform on other modular projects (see case study in Chapter 1 on Baron's Place, Waterloo).

The 3.75 m wide by 12 m long modules comprise a central serviced area with living space on either side. Adjacent modules provide extendable space for the three- and four-bedroom house configurations. In this way, economy of scale in the manufacture of the highly serviced core modules is achieved. The modules are stable and self-supporting and are supported on strip footings. Each single-bedroom module provides a living space of 45 m², which satisfies Lifetime Homes.

From a sustainability point of view, the Harlow modular project achieved an EcoHomes "Excellent" rating (equivalent to the Code for Sustainable Homes level 3). Code level 4 can also be achieved by using ground-sourced heat pumps based on 45 m deep piles, which were included in the affordable housing phase. The second phase of the project has also been completed, which included some five-bedroom houses, and was designed to include a range of optional renewable energy technologies.

CASE STUDY 19: SOCIAL HOUSING, TOWER HAMLETS, LONDON

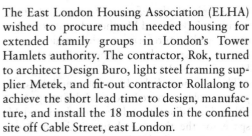

View of completed town houses. (Courtesy of Rollalong and Metek UK Ltd.)

Installation of modules on brickwork footings and strip foundations.

The East London Housing Association (ELHA) wished to procure much needed housing for extended family groups in London's Tower Hamlets authority. The contractor, Rok, turned to architect Design Buro, light steel framing supplier Metek, and fit-out contractor Rollalong to achieve the short lead time to design, manufacture, and install the 18 modules in the confined site off Cable Street, east London.

The project consists of three houses in terrace form and comprises 18 modules. The 9 m long by 3.4 m wide and 12 × 3 m modules are arranged in pairs over 3 storeys. Six modules form one town house and provide spacious accommodation of 183 m² floor area. Each town house had its own integral stairs, patio doors, kitchen, and two bathrooms.

The light steel framework of the modules was manufactured by Metek and transported to Rollalong in Dorset for fit-out and servicing. The fit-out process took approximately 6 weeks, and the modules were then made weatherproof. Six fully finished modules per day were delivered to a holding area not far from the site and brought to the site as required.

The modules were partially open-sided so that two modules form one floor per house. Stair access at the end of alternate modules required the design of a partially open-topped module. The building was traditionally brick-clad with a pitched tiled roof, which were both carried out

off the critical path during finishing work and external service connections.

The whole construction programme took only 20 weeks, a savings of over 50% relative to fully on-site construction. Importantly, there was no long-term disruption to the local area during building, as installation of the modules was carried out in 3 days during the period of 10 a.m. to 3 p.m., agreed with the Tower Hamlets Local Authority.

The modules consist of end wall panels using 100 × 1.6 mm thick C sections and side wall panels using 65 × 1.2 mm C sections. Sides are braced using 100 mm wide cross-flats for stability under wind load. The floors consist of 150 × 1.6 mm C sections, which are placed back-to-back at 400 mm centres for stiffness purposes. The ceilings also use 65 × 1.2 C sections, and the combined depth of the floor and ceiling was only 300 mm. Steel 100 × 75 × 8 mm angles were introduced in the corners of the modules.

The installation period within the agreed 3-day road closure period saved 4 months on the construction programme, and eliminated the need for storage of materials, site huts, and other equipment. A total of four workers were involved in the installation process, and three workers were engaged for general duties, which was less than 20% of the workforce required in traditional site-based construction.

CASE STUDY 20: CODE LEVEL 5 SOCIAL HOUSING IN WEST LONDON

Completed building with its curved green roof and PV panels. (Courtesy of Futureform.)

Framework of the Ayrframe module. (Courtesy of Ayrshire Framing.)

Birchway Eco-Community is a development by the Paradigm Housing Association of five buildings comprising 24 new homes for affordable rent in Hayes, west London. Modular construction was chosen for this project because it had the minimum environmental impact and it caused the least disturbance to the neighbourhood to the former allotment site.

The project was designed by architect Acanthus, and it was one of the first to satisfy the Code for Sustainable Homes level 5. A range of measures was used, including low-carbon centralised biomass boilers, photovoltaic panels, mechanical ventilation and heat recovery, and rainwater retention and recycling. The biomass boilers are located in an underground chamber and approximately 1 tonne of biomass in the form of wood pellets is consumed annually per apartment, and the delivered hot water is metered in each apartment.

Two building types were designed that comprised 10 or 13 modules in one- and two-bedroom configurations. Two modules of approximately 3.6 m width by 9 m length form a single-bedroom layout, and three modules form a two-bedroom layout. Four or six apartments comprise one building. The Futureform designed

modules were manufactured using the Ayrframe fabrication system, and were fully fitted out in the Futureform factory. A single module comprises up to two bedrooms and a bathroom, or the living and kitchen space. The communal stairs and lobby were also constructed as modules.

The individual buildings are 2 storeys high but have a curved green roof to minimise the visual impact and water runoff. The curved roof was formed by manufacturing the upper floor modules with a variable slope ceiling, which was designed to support the weight of the sedum green roof. The roof also supports PV panels on the south face, which provide electricity for the communal spaces. The double-floor and wall construction also provides excellent acoustic insulation between dwellings.

The fully finished modules were installed at the rate of 8 per day, which meant that the whole installation for one building was completed in 2 days rather than months. This was important for the neighbouring houses in the tight urban site, and installation on the same day each week was carried out by informing the nearby residents. A 100T mobile crane was used because of the long distance over which the modules were lifted and placed on site.

CASE STUDY 21: MODULAR HOUSING FOR PURCHASE, CUB HOUSE

CUB Housing at BRE Innovation Park. (Courtesy of Futureform.)

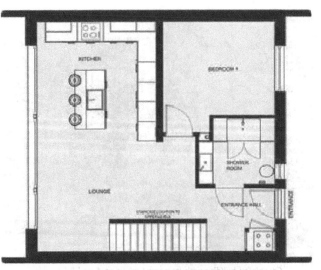

Ground floor plan showing two adjacent modules. (Courtesy of Futureform.)

The CUB Housing range was launched at the Ideal Home Exhibition in March 2010. It is manufactured by Futureform and is based on two open-sided 3.5 m wide by 7 m long modules forming a square living space, which can be extended as the users' needs grow. The 2:1 plan aspect ratio of the modules means that they can be grouped together to form larger spaces. The system is also extendable from houses to multistorey buildings. A variety of cladding materials may be selected, which can be either preattached to the modules or site installed.

The CUB concept was developed by designer Charlie Grieg together with Futureform. Buyers can choose from a wide range of floor, wall, kitchen, and bathroom finishes, and the modules can even come with fitted furniture. The pair of modules comprises a spacious kitchen/living room, bedroom, and shower/toilet. A typical plan form of the ground floor is shown.

The modules are manufactured in light steel framing, based on the Ayrframe system. Modules include argon-gas-sealed windows with low UV values, an exhaust air heat pump in a housed unit, rainwater recycling, and an option for solar panels. Service connections are made on site, and the joint between the modules is sealed. The walls, floor, and ceiling of the modules are manufactured as weather- and airtight, and are highly insulated. Running costs are estimated at £56 per year for heating, hot water, and ventilation.

Estimated delivered costs are £88,500 for the 7 m² double-module configuration, plus £10,000 to £15,000 site preparation costs. The modules take 12 to 14 weeks to manufacture from order and are installed in only a few days, depending on the suitability of the site. This represents a total time savings of over 6 months relative to traditional housing build times.

CUB Housing at a system level satisfies level 5 of the Code for Sustainable Homes, and the system is approved by the National Housing Building Council. The two-module configuration can be extended to four or more modules without changing the basic structure. It is based on the generation house concept that can be modified as family sizes change.

The modules come with a removable ceiling panel so that the stairs can be installed if two additional modules are placed on top of either a two- or four-module configuration. Service connections can be made through accessible service risers. Its fire resistance of at least 60 min means that it can be used in multistorey buildings without changing the basic design. All CUB modular buildings can be easily disassembled, which means, in theory, they can be moved either on the same site or to other locations, and their asset value is maintained.

CASE STUDY 22: MODULAR HOUSING AND ROOFTOP EXTENSIONS, FINLAND

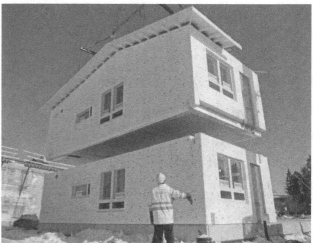

Two-storey sheltered accommodation in Vantaa. (Courtesy of NEAPO.)

Two-storey housing using large modules in Espoo. (Courtesy of NEAPO.)

NEAPO, based near Tampere in Finland, supplies a novel structural system based on light steel wall and floor panels for use in large modular units. The panel system is called Fixcel, which is a honeycomb panel manufactured using thin steel that is folded and compressed at its ends to form a structurally very robust and stiff multiple "box" section. The Fixcel wall panel is 100 mm deep using steel of 0.7 to 1.2 mm thickness, and the floor panel is 150 mm deep. Both panels can be manufactured up to 5 m wide and 22 m long.

The main applications of this modular system are in residential buildings and in specialist applications, such as elevator shafts and modules for rooftop extensions. Modular sizes are limited only by transportation, and modules up to 5 m wide have been supplied on recent projects in Finland. An example of 2-storey sheltered housing in Vantaa using the NEAPO modules is shown above. The insulated render cladding was attached in the factory, and only the joint between the modules was made good on site.

The double-skin walls resist all loads applied to the building, and no separate structure or bracing is required. The same system is used for the floors and ceiling of the modules. The modules are insulated externally by up to 200 mm of rigid insulation board. Fire tests at the Technical

Research Centre (VTT) in Finland confirmed 120 min of fire resistance (132 min by test) with two layers of plasterboard internally.

The modules are manufactured using prefabricated wall, floor, and ceiling panels. Temporary bracing is introduced in the large open sides during transportation. Large openings can be created. A module weighs only 250 to 300 kg/m² floor area, based on a Fixcel panel weight of 17 to 25 kg/m². Fitted-out modules are relatively light, at around 10 tonnes for a typical module of 40 m² floor area.

In Espoo, a series of 37 terraced houses each of 2 storeys was constructed using large modules of 55 m² floor area. The upper modules were manufactured with a pitched roof, and were installed as shown in the figure above. Balconies were attached via inclined tubes.

In an interesting renovation project in Helsinki, two new floors comprising modules were built on top of an existing 5-storey residential building. New lift shafts and stairs were also added at the ends of the building using the same form of construction. The lift shafts were preassembled on site before lifting into place. Also in Helsinki, a floating house called Villa Helmi was created using three modules.

CAST STUDY 23: SOCIAL HOUSING, KINGS CROSS, LONDON

View of the building showing the large balconies attached to the modules.

Internal view of fitted-out module.

This 5- and 6-storey social housing project in Calshot Road, near Kings Cross in north London, was completed in 2006 by Caledonian Modular for the Genesis Housing Association. The L-shaped building consists of 23 apartments, each of two bedrooms, and 9 town houses, each with five bedrooms. The town houses are designed in 3-storey formats with private entry at the ground floor level. Each apartment has a private balcony.

The town houses each comprise three vertically stacked modules, and each module has two rooms and either a bathroom or a kitchen. The total floor area of a town house is approximately 115 m². Stairs were integral to the modules. The living room and kitchen were located at the ground level, giving the building a feel of terraced housing despite its size.

The modules were relatively large, being generally 3.8 m wide and up to 11 m long. They were manufactured with 100 mm square hollow section corner posts and 170 mm deep steel C section edge beams at the floor and ceiling levels, so that they could be manufactured with partially open sides and can also provide a high degree of fenestration. This also assisted in planning the layout of the town houses and apartments.

The modules were designed as weathertight, and in this project, the cladding was mainly in the form of terra-cotta tiles attached to horizontal rails that were pre-fixed to the modules. In other areas, insulated render was placed on the external sheathing board of the modules. Patio doors were also provided in the corner modules. The stair core was also fully glazed.

Stability was provided by the bracing in the light steel walls of the modules and by additional bracing in the stair and lift cores. The prefabricated corner and side balconies were also directly attached to the corner posts and edge beams of the modules. The upper modules also supported the projecting roof structure.

The project consists of 73 modular units and was completed in only 6 months. The modules were fully fitted out before delivery, and only the floor joint between adjacent partially open-sided modules was made good on site. The total construction cost in 2006 was £2.4 million, which was very cost-effective.

CASE STUDY 24: ELEVEN-STOREY APARTMENT BUILDING, BASINGSTOKE

Eleven-storey modular apartment building.

Six-storey modular apartment building.

One of the first residential projects using the Vision modular building system was constructed in Basingstoke. This development consists of three blocks, ranging from 6 to 11 storeys. The architectural design of the building was developed by PRP for Fleming Developments UK, based on the client's master plan by HTA Architects. The client was the Sentinel Housing Association, and modular construction was selected because of its speed of manufacture, and its minimum disturbance to the nearby hospital. The 360 modules were installed from October 2006 to February 2007.

Two modules created a one-bedroom apartment of 48 m² floor area, and three modules created a two-bedroom apartment of 60 m². The 3 and 3.6 m wide modules were arranged on either side of a corridor and were accessed from the stairs and lift, which service six apartments. The various modules comprised a kitchen/dining area, bathroom and master bedroom, and hall and smaller bedroom.

The modules were typically 3 or 3.6 m wide and 7.2 m in length. The module floor consisted of a 150 mm deep concrete floor with PFC steel sections at the perimeter of the floor. The use of a concrete floor provided a high level of acoustic insulation and 120 minutes of fire resistance.

The walls and roof comprise structural hollow sections welded into frames. Balconies were attached to the perimeter PFC sections.

The modules were lifted by mobile crane and were installed at an average rate of eight per day. Overall lateral stability of the buildings was provided by the reinforced concrete stair/lift cores. For the 11-storey building, the modules were installed in just 15 days, which meant that cladding and follow-on trades could commence immediately. It was estimated that modular technology saved 70% on the project time period relative to in situ concrete construction.

Modules varied in shape from rectangular elements to irregular shapes, with splayed corners and stepping of the walls on plan. The combined floor and ceiling depth was only 350 mm and comprises a ceiling truss, which allows for passage of services. The combined width of adjacent walls was only 200 mm.

The cladding consists of ground-supported brickwork for three or four floors, and insulated render and lightweight panels above. The separating wall between modules achieved an average airborne sound reduction of 52 dB (with the low-frequency correction factor), which is 7 dB better than Part E of the Building Regulations.

CASE STUDY 25: HIGH-QUALITY HOUSING IN CENTRAL CROYDON

View of the front of 6-storey building from the access bridge.

A 6-storey mixed retail and residential development of over 300 modules in the centre of Croydon was completed by Caledonian Modular in 2010. The difficult L-shaped site was hemmed in by the busy Surrey Street with its daily open market and by an existing car park at the rear. The development created much needed private, part-buy, and social housing, primarily in two-bedroom formats, but with some single-person accommodation.

The client's motivation to use modular construction was driven by the difficulties posed by more traditional construction relating to the time and disturbance involved in on-site activities, such as the many deliveries of building materials and the number of construction workers involved. As an example of the constraints imposed by the local authority for this sensitive project, all the modules had to be delivered after normal working hours, and a 500-tonne mobile crane was brought in to lift the 10-tonne modules at a distance of up to 30 m from the roadside and up to 20 m height.

The overall construction period, including the concrete podium level, was an impressive 35 weeks, which was an estimated 50% savings on conventional on-site construction. The reinforced concrete podium provides retail space below, and the building is entered at the podium level via a new copper-clad bridge over Surrey Street. The individual apartments are accessed by a structural steel walkway at the rear of the building that is linked to two building cores, which house stairs and lifts. The cores were also built in steel in modular form, and an interesting feature was that the core modules below the podium level were installed early in the construction process to facilitate access.

Two modules of 3.6 m width and 9 m length formed one two-bedroom apartment. Kitchens and bathrooms were located at the rear of the building to facilitate service access and maintenance from the walkway. Brickwork cladding was used on the street-side to give the building a more traditional appearance, and insulated render was used on the rear side. The 13.5 m high brickwork was supported on the podium level and tied into the modules for stability. Balconies were introduced in the manufacture of most street-facing and corner apartments, and were connected to the corner posts of the modules.

The modules on the top floor were set back from the building edge to provide a continuous patio with a glass balustrade that was attached to the modules below. The gently curved lightweight steel roof was also designed as part of the modular system, and it projected over the patio and walkways.

CASE STUDY 26: STUDENT RESIDENCES, NORTH LONDON

Newington Court, north London.

Woodland Court, north London.

Unite Modular Systems has constructed many student residences in modular construction in London and other cities. At its peak, up to 3000 modules per year were manufactured from its factory based in Stroud, west of England. Some more recent student residence projects in north and east London are summarised as follows:

Newington Court is a 6-storey student residence that consists of two blocks, one in brickwork and the other in insulated render with a featured stainless steel exterior. It was completed in early 2010. The 435 modules create 87 apartments in a four-bedroom format, each apartment with a communal kitchen. It was designed by architect Stock Woolstencroft, and the contractor was Mansell. The contract value was £6.9 million.

Woodland Court consists of 669 modules for 481 en suite bedrooms, 45 kitchens, and 2 apartments. Many modules have bay windows and cantilever over the modules below. Cladding is insulated render and simulated brickwork attached to a separate light steel subframe. A Victorian building was retained on the site, and the 8-storey modular building was confined on three of its sides. The modular part of the project was valued at £8.1 million, and the estimated total cost of the project was £15 million. Modules were installed in 17 weeks at a rate of eight per day. Construction started in mid-2009 and was completed in September 2010. The architect was Hadfield Caulkwell Davidson, and the contractor was the RG Group.

Wedgewood Court is on Holloway Road in north London. It consists of 413 modules for 195 en suite bedrooms in one- or two-module configurations and 39 communal kitchens. Modules ranged up to 4.1 × 6.7 m in size, and all were installed in 57 days. The site was located alongside the main Kings Cross railway line, which posed additional constraints. The modular part of the project was valued at £4.9 million, and the total cost of the project was approximately £9 million. The construction period was from December 2009 to the end of August 2010. The architect was Stride Treglown, and the contractor was Woolf Construction.

Somerset Court near Kings Cross is a student residence built over a primary school. It consists of 190 modules for 168 bedrooms and 22 communal kitchens. The modules were constructed on a first floor podium, and the ground floor provided for new school facilities on the same site. The construction had to be carried out without disrupting the school activities. Modules were installed in only 15 days. The modular part of the project was valued at £1.6 million. The architect was Stride Treglown, and the contractor was Mansell.

Blythedale Court in Mile End, east London, is 7 to 12 storeys high and consists of 309 studio bedrooms built over a communal area at the ground floor constructed in reinforced concrete. It was completed in September 2009. The architect was DMWR, and the contractor was Mansell.

CASE STUDY 27: HIGH-RISE MODULAR BUILDING IN WEMBLEY, LONDON

Installation of lightweight cladding by mast climbers. (Courtesy of Futureform.)

View of completed cladding on 17-storey building. (Courtesy of Futureform.)

Futureform has completed its tallest modular building at 17 storeys in the sight of Wembley Stadium for student housing developer Victoria Hall. The architect was O'Connell East, and the management contractor was MACE. This student residence project was a first in terms of the size of the modules, which were 16 m long and up to 3.8 m wide. In this way, the modules comprised two rooms and a twin corridor, which minimised on-site work. The modules were delivered with additional plasterboards so that the corridors could be finished after service connections had been made along the building.

The building consists of a concrete core and circular concrete floor plan with north, east, and west modular wings radiating from it. The west wing consists of 16 storeys of modules built on a concrete podium level, and the north and east wings consist of 4 and 6 storeys, respectively, of modules built on a 2-storey concrete podium.

The construction of the cores and podium started in July 2010, and the modules were installed over a 15-week period. In this way, the construction of the cores and installation of the modules could be carried out in parallel. Each wing consisted of 10 modules per floor, which enabled three floors to be installed per week. The project was completed in August 2011.

The study bedrooms were 2.7 m in external width, and the 8 m long kitchens were of 3.8 m width. Some modules are manufactured with chamfered side walls. A cluster of 10 study bedrooms and two kitchens comprised each wing of the building. A typical 16 m long kitchen module weighed up to 12 tonnes. Lifting of the modules was by a 200-tonne mobile crane located on the roadside. Mast climbers provided perimeter protection during installation of the modules and were connected to the lifting points on the sides of the modules.

The lightweight cladding is a rain screen system supported on horizontal rails attached to the modules. The modules are fully insulated and weathertight, and achieved a U-value of 0.21 W/m²K. The mast climbers were also used to install the lightweight rain screen cladding.

The required 120 min fire resistance was achieved by 25 mm thick core board and 15 mm thick fire-resistant plasterboard coupled with mineral wool placed between the C sections in the walls and ceiling of the Ayrframe fabricated modules. The floor and ceiling joists consist of 150 mm deep C sections, and the combined floor and ceiling depth was only 380 mm. A resilient strip between the modules reduced acoustic transmission and allowed for construction tolerances.

REFERENCES

British Standards Institution. (1990). *Fire precautions in the design, construction and use of buildings: Code of practice for residential buildings.* BS 5588-1.

British Standards Institution. (2008). *Code of practice for fire safety in the design, management and use of buildings.* BS 9999.

Building Regulations (England and Wales). (2010a). *Fire safety. Volume 1. Dwelling houses.* Approved document B. www.planningportal.gov.uk.

Building Regulations (England and Wales). (2010b). *Resistance to passage of sound.* Approved document E.

Building Regulations (England and Wales). (2010c). *Conservation of fuel and power.* Approved document L1.

Building Regulations (England and Wales). (2013a). *Fire safety. Volume 2. Buildings other than dwelling houses.* Approved Document B.

Building Regulations (England and Wales). (2013b). *Access to and use of buildings.* Approved document M.

Code for Sustainable Homes. (2010). Technical guidance. www.gov.uk/government/publications.

Housing Corporation. (2007). Design and quality standards.

Joseph Rowntree Foundation. (2010). Lifetime homes.

Mayor of London. (2010). Design for London, London housing design guide—Interim. www.designforlondon.gov.uk.

National Housing Federation. (2008). *Standards and quality in development: A good practice guide.* 2nd ed.

Secured by Design. www.securedbydesign.com.

Zero Carbon Hub. (2009). *Defining a fabric energy efficiency standard for new homes.* www.zerocarbonhub.org.

Chapter 7

Hospitals and medical buildings

Hospitals and healthcare facilities are often constructed using modular construction, either for specialist rooms or as complete modular buildings. The main advantage of modular construction is that the installation of complex services and fitting out of medical equipment can be carried out and checked in clean controlled factory conditions.

Modular construction can be used to provide complete new hospitals or to extend existing healthcare buildings, in which modules are installed with minimum disruption to patient care. In addition, specialist modular facilities are wards, operating theatres, diagnostic imaging suites, laboratories, mortuaries, decontamination units, and plant rooms.

7.1 FEATURES OF MODULAR MEDICAL FACILITIES

There are various providers of modular medical buildings, and their modular systems conform broadly to the following characteristics:

- Single or multistorey applications (up to 6 storeys high)
- Clear useable space of up to 12 m between columns
- A range of internal finishes and exterior cladding options
- Internal layouts based on critical pathways through the buildings
- High degree of pre-installed services and medical equipment

Modular units for the use in the health sector are often relatively large and have partially or fully open sides. In this way, the various types of functional spaces that are provided in modular form are combined to form complete healthcare buildings. Although the size and structure of the modules may be similar, the individual modules are manufactured with their specialist equipment and services, depending on their functional use. Importantly, these installations are tested and commissioned before delivery to the site.

Modular construction is often combined with a primary structure for the open-plan, circulation, and entrance areas. Examples of recently completed hospitals in modular form are illustrated in Figures 7.1 and 7.2.

Modules used in the medical sector generally have deep edge beams and corner posts so that clear spans of up to 12 m can be created. The edge beams in the ceiling of the module generally project upward rather than creating down-stand beams between the modules. The combined depth of the edge beams in the floor and ceiling can be as high as 750 mm. Module widths may vary depending on the layout of the medical space and its specialist facilities.

Yorkon supplies modules in 3.75 m width and in lengths in increments up to 18.75 m that are used in this and other sectors. Intermediate walls and posts may be provided as part of the otherwise open-sided modules. At the Colchester General Hospital, shown in Figure 1.8, a new building accommodates a new children's department on the ground floor and an elective care centre and surgical ward on the upper floor. A total of 148 steel-framed modules up to 14 m long and 3.3 m wide, each weighing up to 12 tonnes, were delivered and installed in just 17 days. In this project, the units were partially fitted out off site, including internal partitioning and first fix mechanical and electrical (M&E) services.

The module sizes used in wards are typically 3.6 to 4 m wide and 7.5 to 18.75 m long. They are generally partially open-sided and often include corridor space. Often intermediate walls and posts are provided, as shown in Figure 7.3. A typical plan form of a large ward consisting of two and three bedrooms is shown in Figure 7.4. Specialist rooms, changing rooms, and bathrooms are also provided within the modules.

Where there are sufficient internal walls, overall stability of the building may be provided by the modules themselves, but where there are few internal walls, stability has to be provided by additional bracing around the stair and lift cores. The normal building height for this type of open-sided module is typically 3 storeys, but medical buildings up to 6 storeys high have been designed.

Figure 7.1 Completed 3-storey hospital in modular construction in Bristol. (Courtesy of Yorkon.)

Figure 7.2 Completed NHS Treatment Centre in Portsmouth in modular construction. (Courtesy of Yorkon.)

New medical facilities are often required within existing hospitals, and so space for on-site construction processes and storage of materials is often limited. In this case, the use of modular construction is beneficial to avoid disruption to the functioning hospital. Modular units are often used in the extensions of existing health-care buildings, for example, by adding new floors (see Section 7.3).

Modular construction may also be used for local medical centres, such as in Figure 7.5, where an 800 m² general practitioner (GP) centre in Hillingdon, west London, was built in only 6 months.

Figure 7.3 Modular hospital unit during construction. (Courtesy of Yorkon.)

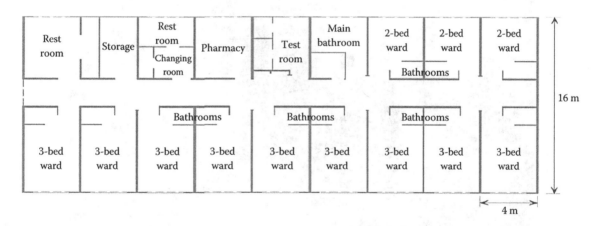

Figure 7.4 Example of layout of modular rooms in a hospital ward. (Courtesy of Cadolto.)

7.2 DESIGN REQUIREMENTS

Comprehensive design guidance for medical buildings is provided by the UK Department of Health in the series of publications entitled Health Technical Memoranda (HTMs) and Health Building Notes (HBNs). The HTMs cover subjects such as decontamination, sustainable planning, design, construction, and refurbishment of health and social care buildings, design and management of building services and equipment, fire safety, ventilation, and transport management. The HBNs provide planning and design guidance for a wide range of medical facilities, finishes, sanitary equipment, infection control, resilience planning of facilities, circulation and communication, and management of land

and property. The following sections consider a few of the main design considerations for healthcare buildings.

7.2.1 Acoustics

HTM 08-01 sets out the recommended acoustic criteria for the design and management of new healthcare facilities, covering issues such as

- Noise levels in rooms, including contributions from both mechanical services within the building and external sources transmitted via the building structure
- External noise levels—noise created by the healthcare building and operation should not affect those that live and work around it

Figure 7.5 Local medical centre in modular form. (Courtesy of Elliott Group Ltd.)

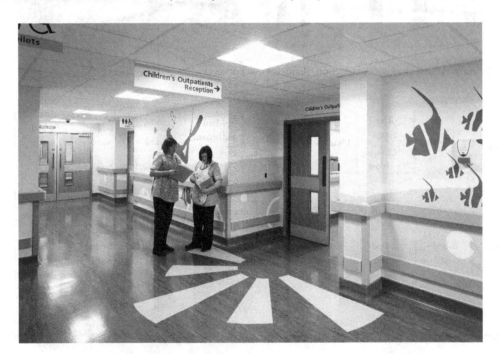

Figure 7.6 Circulation areas in a modular hospital in Colchester. (Courtesy of Yorkon.)

- Airborne and impact sound insulation between rooms
- Control of reverberation in rooms

The Department for Health has also published *Acoustics: Technical Design Manual 4032*, the content of which is very similar to that of HTM 08-01.

7.2.2 Circulation areas

The document HBN 00-04 provides guidance on the design of circulation and communication spaces in hospitals and other healthcare buildings, including corridors, internal lobbies and stairs, and lifts. It also provides supporting information on doors and handrails.

The main corridor and circulation space is often up to 3 m wide and is provided as part of the open-sided module. An example is shown in Figure 7.6. Stairs are also generally provided in modular form, and their width is dependent on the requirements for means of escape in fire (see Section 7.4).

7.2.3 Wards

Wards can include large multisection wards, acute wards, isolation rooms, single-bed rooms with en suite, intensive care units, and recovery areas. An example is shown in Figure 7.7. Wards in modular construction are fitted with the following:

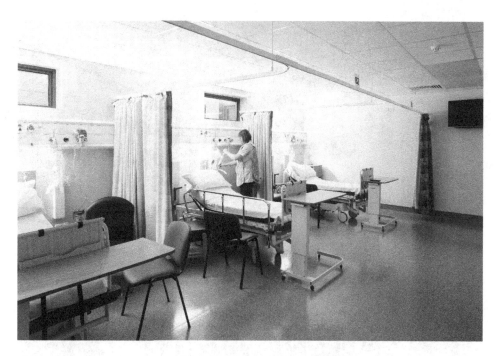

Figure 7.7 Typical open-plan ward in modular construction. (Courtesy of Yorkon.)

- Easy-to-clean and easy-to-maintain finishes and surfaces
- Comprehensive bedhead services, including telecommunication systems
- Medical gases and air handling systems in recovery and high-dependency units
- Nurse stations, equipped with alarm and monitoring systems
- A floor structure designed to improve the sound absorption of the structure

7.2.4 Sanitary spaces

Sanitary spaces include bathrooms, changing rooms, and toilets, which should satisfy HBN 00-02. This covers spatial requirements and ergonomics, including wheelchair access.

7.2.5 Specialist facilities

Specialist medical buildings generally require special facilities that are complex to install, for which off-site manufacture provides the ability to pretest and commission before delivery. One of Scotland's first birthing suites opened in October 2011 and was constructed in modular units. The completed building is shown in Figure 7.8.

At the Nottingham University Hospital's NHS Trust, a new renal unit was completed by Portakabin in modular form that included 10 dialysis stations. Renal healthcare buildings should satisfy HBN 07-02 requirements. The water treatment plant, services, and internal finishes are preinstalled in the modules.

Specialist consultation rooms can be provided in a wide range of medical facilities, and an example is shown in Figure 7.9. These rooms are generally equal in width to the module but can be linked to the adjacent corridor so that two consultation rooms may be provided in a single module.

7.2.6 Operating theatres

Modular operating theatre suites can be produced for surgical needs in all medical fields, including ophthalmology, orthopaedics, cardiology, neurology, and oncology. In the UK, they must be designed in accordance with HBN 26: *Facilities for Surgical Procedures* and HTM 03-01: *Specialised Ventilation for Healthcare Premises*, and they should also be fit for purpose in terms of the life cycle of the facility.

Operating theatres are usually housed in one or two modular units, which also contain an anaesthetic room, scrub areas, preparation rooms, and utility areas. An example is shown in Figure 7.10. Concrete floors are often required to limit any vibration effects transferred from the adjacent parts of the building. Special facilities, such as an ultra-clean laminar airflow system or floor-mounted microscopes, may be required. Associated services are usually located in a roof-mounted plant room.

7.2.7 Diagnostic imaging suites

The design of diagnostic imaging suites depends on equipment selection and various technical issues, such as

Figure 7.8 Birthing suite facility in Lothian, Scotland. (Courtesy of BW Industries.)

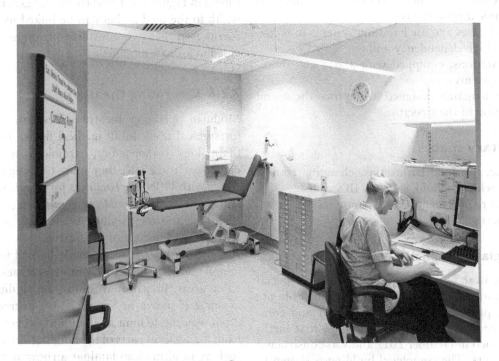

Figure 7.9 Typical consultation room within a larger module. (Courtesy of Yorkon.)

magnetic field effects, interference, magnetic shielding requirements, co-siting issues, fringe field distribution, radio frequency (RF) enclosures, and equipment installation. Pressure equalisation, oxygen monitoring systems are required.

Modules can be manufactured with concrete floors to support heavy equipment and to suppress vibration effects in specialist applications, such as diagnostic imaging suites and surgical theatres. The concrete may be placed as a thin topping to a joisted floor, or

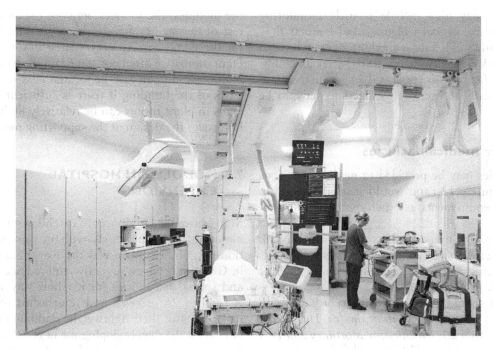

Figure 7.10 Operating theatre using two modules. (Courtesy of Yorkon.)

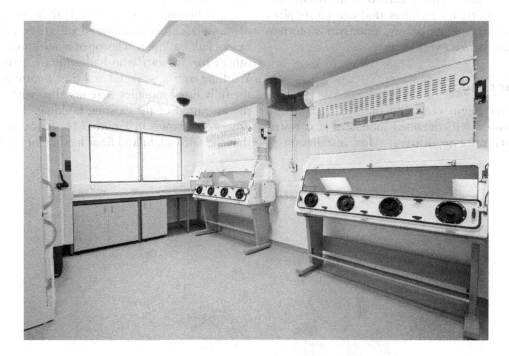

Figure 7.11 Laboratory constructed using modular units. (Courtesy of Yorkon.)

as a solid slab cast between the edge beams of the floor of the module. The design should also comply with HBN 6: *Facilities for Diagnostic Imaging and Interventional Radiology.*

7.2.8 Laboratories and clean rooms

Modular pathology laboratories, pharmaceutical aseptic suites, containment laboratories, etc., should comply with the latest guidance from the Advisory Committee on Dangerous Pathogens standards (Health Protection Agency) and associated clean rooms with ISO 14644 (1999) design standards (ISO 2000, 2001, 2004, 2005).

An example of a modular laboratory with specialist facilities for handling pathogens is shown in Figure 7.11. This type of module requires a controlled clean air environment that is provided by supply and extract air filtration and an integrated fumigation system. This may also include particle monitoring facilities, pharmaceutical isolators, and laminar flow cabinets. A fully welded sheet vinyl interior can be used to provide leak integrity that is tested after assembly. Most of these facilities are preinstalled within the modules, so that only the service connections have to be made on site.

7.2.9 Other medical facilities

Dental facilities may be provided in modular form, as shown in Figure 7.12, and would generally conform to the same requirements as other specialist consultation rooms.

Decontamination units should comply with HBN 13: *Sterile Services Department*. Internal finish materials must be hygienic and easily cleaned, such as vinyl or stainless steel.

Mortuaries should be designed in accordance with HTM 03-01: *Specialised Ventilation for Healthcare Premises* and HBN 20: *Facilities for Mortuary and Post Mortem Room Services*. The internal finishes must be exceptionally hygienic and be able to be steam cleaned. The range of mortuary facilities that can be supplied includes body storage, pathology, autopsy/postmortem suites, and disaster/recovery facilities.

7.2.10 Plant rooms

Modular plant rooms centralise the M&E plant and equipment into a self-contained module that is tested in the factory prior to delivery. After installation of

the modules, the services between the module and the rest of the facility are connected on site. The controls for the whole facility are situated in one secure area. Plant rooms are often positioned on the roof of modular buildings, but can also be located next to the working parts of the building if there is sufficient space. Floor loads in plant rooms can be relatively high, which also influences the design of the supporting modules.

7.3 MODULES IN HOSPITAL EXTENSIONS

Many hospital buildings are extended while having to maintain their operations without significant disruption. A good example was the Royal Surrey Hospital in Guildford, which urgently required accommodation and specialist facilities for its short-stay surgery unit. A total of 26 recycled and refurbished modules were installed with their lightweight cladding on brickwork plinths on the steeply sloping site next to the busy A and E reception. The completed single-storey building is shown in Figure 7.13.

For a hospital extension in Harrogate, a new floor of modular units was placed on an existing 2-storey brick-clad building. The open-sided modules provide offices for 50 staff who had separate external access. Working over a weekend and using a 500-tonne crane, the fully fitted modules were installed without affecting the operation of the existing building. The installation process is shown in Figure 7.14. It was also assessed that the self-weight and floor loading of the lightweight

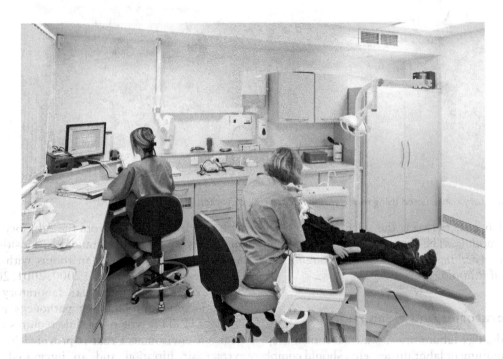

Figure 7.12 Specialist dental facility constructed using refurbished modules. (Courtesy of Foremans Relocatable Building Systems.)

Figure 7.13 Modular construction at the Royal Surrey Hospital, Guildford. (Courtesy of Foremans Relocatable Building Systems.)

Figure 7.14 Installation of open-sided rooftop modules at Harrogate District Hospital. (Courtesy of Yorkon.)

modules did not overload the structure and foundations of the existing building.

7.4 DIMENSIONAL REQUIREMENTS FOR MEDICAL BUILDINGS

Hospitals and medical buildings in modular construction should be designed to satisfy the specified dimensional requirements by using similar-sized modules to form the individual rooms and specialist facilities. Therefore, the early involvement of the modular supplier is essential to optimise the building layout in modular form.

The preferred dimensions are based on a 1200 mm grid, or as fractions, i.e., 600 mm. A modular grid of 3.75 or 4.2 m is commonly used in the heath sector.

The recommended dimensions for the rooms in medical facilities are very detailed, and an indication of the typical minimum requirements is presented below.

Often, they do not conform to the planning grid that may be preferred for repetitive use of modular units.

- Corridor width—Main corridor width of 3 m, or 2.25 m in the ward/working areas and 2.3 m in intensive care facilities.
- Lift lobby—4.7 m wide, bed lift internal shaft 2.4 × 3.0 m.
- Operating theatre—6.5 × 6.5 m on plan and 3 m high, plus 700 mm ceiling depth to accommodate air conditioning and other services.
- Anaesthetics room—3.8 × 3.8 m on plan.
- Washroom—Minimum of 1.8 m width.
- Recovery room/intensive care room—7.2 × 3 m (for two beds).
- Care room—3.9 × 3.3 m (one bed) or 8 m² per bed for multibed rooms.

- Patient shower room—1.4 × 1.4m.
- Patient bathroom—3.5 × 4.35 m.
- Radiology/radiotherapy—Consider weight of equipment (up to 14 tonnes) and structural shielding (generally by lead inserts or thick concrete walls).
- Laboratories should be large enough to offer a high degree of flexibility in use.

Stairs should conform to the requirements of the Building Regulations. These standards cover issues such as stair pitch, width, headroom, length, dimensions of landings, balustrades, fireproofing, etc. Stair widths are determined by escape in case of fire. For example, stairs in public buildings should be a minimum of 1 m wide, or 1.1 m for assembly buildings. Wider stairs (up to 1.8 m) are required for special cases of evacuation.

CASE STUDY 28: MODULAR HEALTHCARE IN BRISTOL, COLCHESTER, AND STOCKTON

Entrance area to Emersons Green Hospital, Bristol. (Courtesy of Yorkon.)

Internal views of the reception area of the Stockton Hospital. (Courtesy of Yorkon.)

The Emersons Green NHS Treatment Centre in Bristol is a £15 million surgical hospital that provides a range of services, including 4 operating theatres, diagnostic and x-ray rooms, 33 beds and reception, and café and administration areas. The 3-storey 4840 m² building comprises 114 steel modules up to 14 m long, which were installed on site in just 3 weeks.

Modules were supplied by Yorkon for healthcare providers' UK Specialist Hospitals (UKSH), and the project was designed by architect TP Bennett. The overall construction time for the project was just 8 months, and parts of the building were handed over early to facilitate fit-out.

The building is clad in cedar boarding with areas of terra-cotta, render, and aluminium rain screen cladding to contrast with the glazed atrium. Concrete floors were built into the modules used in the operating theatres to accommodate highly vibration-sensitive medical equipment. Sustainability features service monitoring, water conservation devices, and solar shading, as well as a "green" roof on the adjacent service building. A material recycling rate of 92% and a 90% reduction in vehicle movements to the hospital site were achieved.

Yorkon has also completed a £20 million hospital building in Colchester under ProCure 21 for contractor Kier Construction. The architect was Tangram Associates. The modular part of this project was valued at approximately £10 million.

All 148 modules were installed in 17 days, and the overall construction programme was only 7 months—an estimated savings of 45%.

The 3-storey fully modular building provides 70 beds, a care centre, consultation and treatment rooms, surgical wards, offices, a school room, dining room, toilets, stores, etc. The modules were 14 m long and 3.3 m wide, each weighing up to 12 tonnes. This project also used preinstalled concrete floors in the modules throughout the building.

The articulated façade consists of insulated render, rain screen cladding, and curtain walling. Modules are also arranged around a courtyard that was part of the architectural concept. The building had one of the most complex layouts designed in modular construction. A key feature was the design of four-bed bays, with each bed having a window.

Another example of the use of Yorkon's off-site building system was at the University Hospital of North Tees in Stockton. This £2.8 million ProCure 21 project with Interserve Project Services included both the construction and fitting out of a 42-bed emergency assessment unit. This 1710 m² single-storey extension comprised 42 modules, which were installed on site in a few days, thus minimising disruption to patient care and dramatically reducing the programme time for the opening of the new facility.

CASE STUDY 29: LEWISHAM HOSPITAL USING TIMBER AND STEEL MODULES

Modules used to extend the Lewisham Hospital.

A new outpatient suite was built using a mix of timber and steel modular construction as part of a major hospital extension at Lewisham Hospital in south London, for contractor Kier London. The hospital had very restricted site access between adjacent medical buildings. The choice of modular construction meant that the hospital building could be constructed in the shortest time and with the minimum site deliveries and noise.

The 3-storey building was specified using nonstandard module widths manufactured by Terrapin. There were a number of stringent aesthetics, performance, and environmental criteria that this project had to meet. The building was designed to achieve BREEAM "Very Good," which meant that it had to meet high performance requirements for airtightness and energy efficiency. Other environmental features included a sedum green roof.

One requirement was to ensure minimal reverberation between the floors adjacent to the corridors, for which an innovative solution involved reinforced high-strength floors to reduce any sound transfer or vibration. A sophisticated integrated plumbing system (IPS) was designed with a frame assembly to avoid the need for on-site assembly, thereby speeding up installation on site.

Cladding to the external façade was selected to match the finish of the existing structure. The old and new connected buildings feature the same cladding. The building was also designed with a "man safe" on the roof, which provides a safe area in which maintenance work could be carried out.

REFERENCES

Building Regulations (England and Wales). (2013). Approved document M—Access to and use of buildings. In *Design considerations*. Hospital Technical Memorandum (HTM) 2045. Department of Health, Stationary Office, Milton Keynes, UK. (Largely replaced by later HTMs.)

Department of Health. (2001). *Facilities for diagnostic imaging and interventional radiology*. Health Building Note (HBN) 6. Stationary Office, Milton Keynes, UK.

Department of Health. (2004a). *Facilities for surgical procedures*. Health Building Note (HBN) 26. Stationary Office, Milton Keynes, UK.

Department of Health. (2004b). *Sterile services department*. Health Building Note (HBN) 13. Stationary Office, Milton Keynes, UK.

Department of Health. (2005). *Facilities for mortuary and post mortem room services*. Health Building Note (HBN) 20. Stationary Office, Milton Keynes, UK.

Department of Health. (2007). *Specialised ventilation for healthcare premises. Part A. Design and validation*. Health Technical Memorandum (HTM) 03-01. Stationary Office, Milton Keynes, UK.

Department of Health. (2008). *Acoustics*. Health Technical Memorandum (HTM) 08-01. Stationary Office, Milton Keynes, UK.

Department of Health. (2012). *Acoustics: Technical Design Manual 4032*. Stationary Office, Milton Keynes, UK.

Department of Health. (2013a). *Sanitary spaces*. Health Building Note (HBN) 00-02. Stationary Office, Milton Keynes, UK.

Department of Health. (2013b). *Main renal unit*. Health Building Note (HBN) 07-02. Stationary Office, Milton Keynes, UK.

Health Protection Agency, Advisory Committee on Dangerous Pathogens. www.hpa.org.uk.

Health and Safety Executive. (2013). Advisory Committee on Dangerous Pathogens (ACDP). http://www.hse.gov.uk/aboutus/meetings/committees/acdp/index.htm.

ISO. (1999). *Cleanrooms and associated controlled environments. Part 1. Classification of air cleanliness*. ISO 14644-1. Geneva.

ISO. (2000). *Cleanrooms and associated controlled environments. Part 2. Specifications for testing and monitoring to prove continued compliance with ISO 14644-1*. ISO 14644-2. Geneva.

ISO. (2001). *Cleanrooms and associated controlled environments. Part 4. Design, construction and start-up*. ISO 14644-4. Geneva.

ISO. (2004). *Cleanrooms and associated controlled environments. Part 5. Operations*. ISO 14644-5. Geneva.

ISO. (2005). *Cleanrooms and associated controlled environments. Part 3. Test methods*. ISO 14644-3. Geneva.

Chapter 8

Schools and educational buildings

In the educational sector, there has been a long tradition of using modular classrooms as a means to cater for short-term expansion of student populations, and some units are in use well beyond their intended life span. Single-storey portable units are still a popular option for short-term accommodation.

However, modules are increasingly used to construct permanent educational buildings, which are reviewed in this chapter. In particular, the dimensional requirements for modules in educational buildings are specific to this sector.

8.1 FEATURES OF MODULAR EDUCATIONAL BUILDINGS

Module configurations and layouts can be designed to suit individual requirements, but certain common features apply in the educational sector. Open-sided modules can be built in groups of up to four wide to form larger classrooms. However, where open-sided modules are used in buildings over 2 storeys high, the stability of the group of modules has to be provided by a separate bracing system. In schools, the vertical bracing can be located next to stairs and in the end gables.

Yorkon is one of the major suppliers of purpose-designed educational buildings in the UK. Customers can choose from a selection of module sizes ranging in length increments from 6 to 18.75 m. Modular units are manufactured in a standard internal width of 3 or 3.75 m, but non-standard sizes can be manufactured and supplied to order. A variety of site-installed cladding systems or standard factory-installed cladding may be attached to the modules, as shown in Figure 8.1.

Often the modular part of the school building is combined with open-plan areas for the entrance and circulation space that are built in steelwork. For some modular suppliers, this may be delivered as part of the modular package. A good example of the use of a steel structure for the entrance area of an otherwise fully modular building is shown in Figure 8.2.

Some examples of modular school buildings are as follows: The extension to York High School comprised

52 modular units in two different sizes, which were craned into position in just 6 days, reducing disruption to the daily operation of the school. The 2-storey 1900 m² building replaced a number of existing buildings and accommodated specialist facilities. It was completed in only 6 months, an estimated savings of up to 6 months on traditional construction.

A sixth form college near Wigan required 13 new classrooms, which was built during the summer vacation using modules supplied by Foremans. The completed building is shown in Figure 8.3. Solar shading was bolted to the corners of the module for additional strength. A new sixth form centre at the Alperton Community School in Brent, northwest London, was constructed entirely of refurbished modular units. The 2-storey scheme used 46 modules. Other features of the scheme included preinstalled timber cladding, energy-efficient lighting, solar shading, and water conservation devices.

A new 2-storey library and other specialist teaching rooms were built at the Hayes school in Bromley, Kent, by Elliott Modular using open-sided modules. The building was clad in insulated render and cedar wood and was designed with a roof overhang for shading. External and internal views are shown in Figure 8.4.

CABE (2010) has also promoted better design in schools by providing information to clients. Improvements in school toilets were a part of the former Building Schools for the Future (BSF) programme. Modular toilet blocks have been provided by various suppliers in accordance with the Department for Children, Schools, and Families (DCSF) *Standard Specifications, Layouts and Dimensions* (SSLD) for toilets in schools.

Modular school buildings may be constructed using light steel modules, but Terrapin has developed its timber Unitrex system for this application, as described in Chapter 4. An example of a single-storey primary school using this system is shown in Figure 8.5. The layout of seven modules to form a single-storey school building for temporary or permanent use is shown in Figure 8.6. In this example, the modules are 8 m long internally and 3 m wide.

Figure 8.1 School building constructed using modules and with site-installed cladding. (Courtesy of Yorkon.)

Figure 8.2 Mixed use of steel frames and modules in a school entrance area. (Courtesy of Yorkon.)

Figure 8.3 Modules with prefinished cladding used in Sixth Form College near Wigan. (Courtesy of Foremans Relocatable Building Systems.)

Figure 8.4 Example of open-sided modules used in a school building. (Courtesy of Elliott Group Ltd.)

Figure 8.5 Primary school using timber modules and cladding. (Courtesy of Terrapin.)

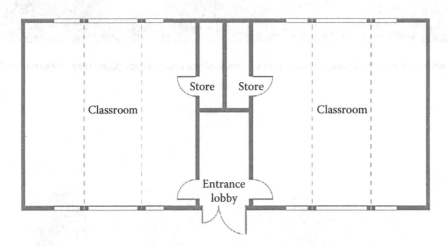

Figure 8.6 Plan form of a group of modules in a single-storey primary school. (Courtesy of Terrapin.)

8.2 DIMENSIONAL REQUIREMENTS FOR SCHOOLS

Guidance on the dimensional planning of secondary and primary schools in the UK is given in Briefing Bulletins 98 and 99 by the Department for Education and Skills (2004a, 2004b). The following important dimensions should be adopted in the planning of modular school buildings:

- Open plan schools are based on a 1200 × 1200 mm grid with a clear internal room height of 3 m for daylight and natural ventilation.
- Primary schools—Classrooms are preferably square and of 63 to 70 m² net floor area, depend-

ing on class size, which equates to 8.4 × 8.4 m for 32 pupils (i.e., 2.2 m² per pupil).
- Secondary schools—Classroom sizes vary with teaching subject and class size. A standard classroom is 60 m² net floor area, and the classroom size increases to 77 m² for an Information and Communications Technology (ICT)-dedicated room or language laboratory. Space requirements are equivalent to 2.0 m² per pupil for traditional classroom teaching, and up to 3.0 m² per pupil for more specialist use.
- Science areas or practical rooms in secondary schools are typically of 90 to 105 m² floor area. The space allowed for a library/media centre is equivalent to 0.35 to 0.55 m² per pupil in the school.

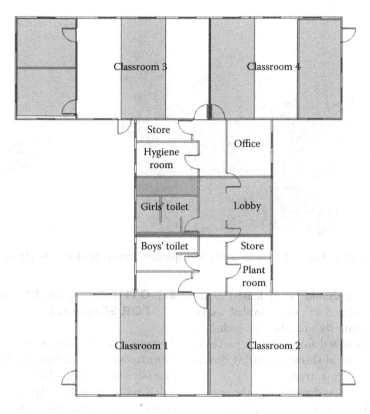

Figure 8.7 Layout of a group of modules to form various classrooms. (Courtesy of BW Industries.)

- Secondary schools require more flexibility in space use, and a typical room size is either 10 m² or 10 × 12 m.
- One toilet for up to 20 pupils and 10 full-time staff, plus disabled facilities, is required, which represents a floor area equivalent to 4 to 7% of the teaching area.
- Dining rooms and sports gymnasia generally require larger and taller spaces.

For design in modular construction, three 2.8 m wide open-sided modules of 8.4 m span are the optimum for primary schools, and three or four 3.3 m wide modules of 10 m span are the optimum for secondary schools. The internal height of the module should be 3 m, which means that the external module height will be 3.6 to 3.9 m, depending on the combined floor and ceiling depth.

An example of the layout of 16 open-sided modules to form four classrooms, together with toilets, stores, and offices of a single-storey school, is shown in Figure 8.7. As noted above, three modules form one classroom, and three similar-sized modules form the toilets, offices, and lobby area. The internal walls are indicated in this figure.

In general, around 25% of the net area of the school is provided for circulation space. Corridors in classroom areas should be a minimum of 2 m wide, or alternatively, 2.7 m when lockers are provided along one wall. Stair widths are determined by escape in case of fire. In schools, the minimum stair width is 1.25 m, but wider stairs (up to 1.8 m) are often required in multistorey schools for rapid evacuation. The rise/tread dimensions for stairs should be typically 170/300 mm. The headroom in stairs should be a minimum of 2 m.

An interesting use of a cluster of modules with a gallery and an atrium roof is shown in Figure 8.8 for a sixth form centre in Essex. In this simple and attractive form of structure, the modules also support the gallery and the roof.

The location of toilets and changing rooms should balance the needs for privacy and supervision. Unisex toilet facilities, with full-height toilet cubicles and doors, leading directly off the circulation areas are becoming more popular in primary schools, as they are more easily supervised. Changing rooms with showers should be placed near indoor and outdoor sports activities, and the size of changing areas should cater for half a year group and should have separate facilities for boys and

Figure 8.8 Layout of a group of modules to form a gallery and spacious atrium at Harris Academy, Essex. (Courtesy of Elliott Group Ltd.)

girls. The space allowance should be at least 0.9 m^2 per pupil in the school plus 5 m^2 for disabled users. Changing facilities for staff should also be included. One shower should be provided for every six or seven pupils in the school. Individual shower cubicles should be at least 1.25 m^2 plus a drying area.

Dining and kitchen spaces depend on the provision planned. A typical figure of 0.9 m^2 per pupil is allowed for dining/socialising space for 11- to 16-year-old pupils. This space is often used for other purposes at other times. The size of the kitchen depends on the catering system.

8.3 OTHER REQUIREMENTS FOR SCHOOLS

According to the orientation of classrooms and other rooms, shading facilities might be required to prevent excessive solar gain. Cross-ventilation should be provided where possible. Attention should be paid to the acoustic environment, both soundproofing between rooms and sound absorption within rooms to reduce unwanted reverberation. Local automatic control of heating, lighting, and ventilation is preferable, with manual overrides to cater for exceptional conditions.

CASE STUDY 30: MODULAR EDUCATIONAL BUILDINGS IN WALES AND WORCESTERSHIRE

Christ College, Brecon. (Courtesy of Yorkon.)

Entrance area to Bewdley High School, Worcester. (Courtesy of Yorkon.)

A college extension constructed using Yorkon modules won the Building Schools for the Future Award at the Builder and Engineer Awards in November 2009. The award recognised the project's high-quality design and how it applied construction best practice to an educational project built almost entirely using off-site technologies.

The use of modular construction helped ensure that the new £1.3 million Hubert Jones Science Centre at Christ College in Brecon was completed in just 5 months despite the challenging site conditions and the need to minimise disturbance to staff and pupils. The 16 modules were installed during the school holidays. Two physics and two biology laboratories, a sixth form project room, and laboratory technician's rooms are grouped around a double-height central atrium that functions as an additional teaching area and exhibition space.

As the college campus is located in the Brecon Beacons National Park, the design incorporated local materials such as Welsh sandstone, together with insulated render and local timber cladding. A number of sustainable design measures were also incorporated, such as solar thermal water heating, energy-efficient lighting, natural ventilation, and high levels of thermal insulation in the fabric of the modules.

The modular building was procured in a design and build contract, and was designed by P+HS Architects. It was designed to be flexible in use and adaptable so that it could be reconfigured to meet the school's requirements in the future. The modules were 3.3 m wide internally, and so three modules with open or closed sides created a large classroom. The internal walls of the modules are non-load bearing, and clear internal spans of up to 12 m enabled the effective layout of teaching spaces and laboratories.

Another example of the use of Yorkon's modular system was the Bewdley High School and Sixth Form Centre designed by architects at Worcestershire County Council. It provided much needed space for an additional 360 pupils joining the school. The project was the first example of a modular building to be heated using biomass fuel.

The 2-storey building consisted of 12 classrooms, 2 science laboratories, a creative area, and an administrative centre. A total of 60 modules were installed with clear internal spans of up to 12 m. The total time on site was just 22 weeks, an estimated savings of over 20 weeks. The entrance area to the school constructed in tubular steel and I section beams is shown above. The rain screen cladding uses timber weather boarding.

CASE STUDY 31: SCHOOL EXTENSION IN SEOUL, KOREA

Completed extension school with its curved roof.

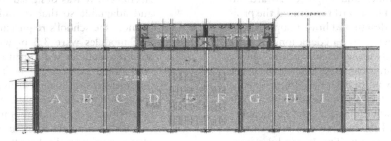

Plan form showing the stair and toilet modules.

The Seoul education authority in Korea wished to extend the provision of school places within existing schools and turned to modular construction for the solution. In 2003, the Shin-Yi Middle School in Seoul city centre was used as a demonstration of a new modular construction system, which was designed by the steel company POSCO with assistance from the Steel Construction Institute and the concept architect, the Design Buro.

The nine classroom modules were manufactured as 12 m long and 3 m wide so that three modules would form one classroom of 9 × 9 m. A 3 m wide corridor was created using an intermediate steel post within the side of the module. The modules used steel channel sections of 300 mm depth as the edge beams in the floor and 200 mm deep channel sections at the ceiling level to span between the 100 mm square hollow section posts. The school consisted of 2 storeys of similar modules, and the overall intermediate floor depth was only 600 mm. Stability to wind loading was provided by the bolted beam-to-post connections.

The four 3.3 m wide and 6 m long toilet modules were manufactured with a concrete floor to facilitate washing down, as is the case in Korea. The enclosure to the stairs was built on site, but in later projects, the stairs were manufactured as modules. The projecting roof was formed using curved light steel panels that were placed on the side walls of the modules, and purlins spanned between these walls. The side walls to the outer modules of a classroom used light steel infill walls that were braced for additional stability. The cladding was in the form of insulated boards that were directly fixed to the infill walls.

The building work took place during the 2-month summer vacation, and the modules were placed accurately on strip footings. The 22 modules were brought through the busy streets of Seoul and were installed over a 3-week period using a 100-tonne mobile crane. Since then, many other school buildings have been built in modular form in Korea using the experience gained on this project. POSCO has also built military barracks up to 4 storeys high using the same modular construction system.

CASE STUDY 32: NEWHAM SIXTH FORM COLLEGE USING TIMBER MODULES

Terrapin provided a fast-track solution to the need for additional learning space at Newham Sixth Form College in East London. The 440 m² single-storey school building was designed and built using panelised timber units from the Unitrex system in a fraction of the time it would have taken using traditional construction methods. Start of on-site work to handover of the finished building took just 4 weeks, with the minimum of disruption to college life. The building's modular units were installed in a single day, and the time from foundation to handover was only 1 month.

The building consists of an entrance lobby, four 38 m² and two 45 m² classrooms, open-plan staff and office accommodation, one-to-one meeting rooms, and storage areas. The façade was finished in western red cedar horizontal cladding, and a variety of window sizes created an interesting façade detail.

Modular educational buildings are available in spans of 4.8 m to 12 m in 1.2 m increments to up to 2000 m² in floor area in 1- or 2-storey configurations. Duopitch roofs may be finished with traditional tiles and monopitch roofs may be finished with lightweight coverings. A selection of partition systems and wall finishes can be selected to suit individual needs, together with electrical and mechanical installations.

REFERENCES

Commission for Architecture and the Built Environment (CABE). (2010). *Creating excellent primary schools. A guide for clients.* London. www.cabe.org.uk/files/creating-excellent-primary-schools.pdf.

Department for Education and Skills. (2004a). *Briefing framework for secondary school projects.* Briefing Bulletin 98 (BB98). Annesley, UK. (Revision of BB82: *Area guidelines for schools.*)

Department for Education and Skills. (2004b). *Briefing framework for primary school projects.* Briefing Bulletin 99 (BB99). Annesley, UK. (Revision of BB82: *Area guidelines for schools.*)

Department for Education and Skills. (2007). *Standard specifications, layouts and dimensions: Toilets in schools.* Nottingham, UK.

Chapter 9

Specialist buildings

A wide range of specialist building types may be designed in modular construction, including:

- Supermarkets
- Retail units, such as convenience stores, motorway services
- Petrol stations
- Military accommodation
- Secure units and prisons
- Airport buildings
- Offices
- Laboratories

All of these buildings types share the common feature that they can be manufactured efficiently off site with a high degree of repetition and quality control, and can be installed rapidly where site logistics require minimum disruption.

9.1 SUPERMARKETS

Yorkon has developed a long-span modular system specifically for Tesco supermarkets, which is based on a 18.75 m length, 3.75 m wide, and 4 m high open-sided module that is designed for single-storey retail applications. This system has been used in remote locations where site logistics are a problem, and on high-value sites where speed of installation and minimising disturbance are important requirements for the project's success.

The edge beams in the modules at the floor and roof levels are typically 350 mm deep. The modules are supported continuously at their base or on pad footings at 3.5 m spacing so that the edge beams can support floor loads up to 5 kN/m². The edge beams at the ceiling level support the roof and service loads over the open span of the module. Intermediate posts can be used to reduce the clear span. Services are installed under the floor of the single-storey module, as shown in Figure 9.1, which is the largest modular supermarket completed to date. In a project in Orkney (see case studies), 26 modules were installed in 3 days and the supermarket was operational just 3 weeks later.

9.2 RETAILS UNITS AND PETROL STATIONS

Kiosks and other small retail units are generally manufactured in modular form, as they are easily transportable and can be moved. Petrol stations are also often constructed in modular form, as they are highly standardised and speed of installation is crucial to the business operation. Figure 9.2 shows a good example of a shop comprising a single module for a petrol station, which is repeated in many hundreds of locations throughout the UK.

The use of modular construction has been well established in fast food restaurants since the early 1990s, but 2-storey modular buildings have been developed for this sector, as illustrated in Figure 9.3 for a recent project in Bognor Regis.

9.3 MILITARY ACCOMMODATIONS

Modular construction has been used for secure accommodation of all types, including Ministry of Defence buildings and military accommodation. This type of accommodation is usually procured on a turnkey or design-and-build basis, including internal facilities and fixed security systems. Off-site construction enables the on-site workforce to be reduced, which minimises the number of security clearances required. Modules can be manufactured in either light steel framing or precast concrete.

The Single Living Accommodation for the Military (SLAM) project has procured 10,000 accommodation units over 8 years, all in modular construction. They were delivered by Aspire, a joint venture organisation that involved various modular suppliers. A typical modular construction of military accommodation is shown in Figure 9.4. One of the first completed military accommodation projects was at Salisbury Plain, which was designed to be BREEAM "Excellent" (see case study).

The WRAP (2008) Woolwich Single Living Accommodation Modernisation Regeneration study reviewed the opportunities to reduce waste and recycle waste in this military accommodation project in

Figure 9.1 Single-span open-sided modular supermarket at Southam, Warwicks. (Courtesy of Yorkon.)

Figure 9.2 Petrol station in modular form. (Courtesy of Caledonian Modular.)

Figure 9.3 Fast food restaurant built in fitted-out modular units. (Courtesy of Elliott Group.)

Figure 9.4 Typical 3-storey military accommodation fully in modular construction. (Courtesy of Rollalong.)

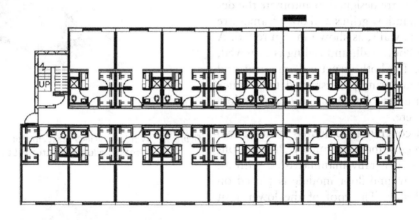

Figure 9.5 Plan view of modules for military barracks. (Courtesy of Oldcastle Precast.)

southeast London. In this project, the modules were manufactured in light steel framing with a thin steel sheet attached to the outer face of the modules to provide resistance to blast fragmentation. The accommodation blocks are generally 2 or 3 storeys high using modules of typically 3.6 m width and 7 m length. Stairs are also constructed as modules, but corridors are manufactured as planar elements. Construction periods were reduced by over 60% relative to traditional building, and importantly, the personnel employed on site were reduced by 70%.

Concrete modules are also used in all types of buildings where security is important. The construction technology is revised in Chapter 3 and by Brooker and Hennessy (2008). An example of a concrete modular system for military barracks is illustrated in Figure 9.5. In this scheme, each 4 m wide module accommodates two military personnel.

9.4 PRISONS AND SECURE ACCOMMODATIONS

The UK government's £1.2 billion prison building programme was aimed at creating an extra 15,000 prison places by 2014. Three large Titan prisons of around 2,500 prisoners each were planned with recommendations that off-site and prefabricated methods of construction should be used in order to reduce cost and increase quality.

There are several companies that produce specialised modules suitable for constructing prison and other secure facilities, as follows.

9.4.1 Steel prison modules

Steel prison modules have been developed that have Ministry of Justice approval for use in category B and C regimes, suitable for use as permanent custodial facilities. A complete modular unit comprises six cells per module. Each cell is 7.2 m² in floor area and is designed for single occupancy with its own shower, hand basin, and toilet facility. A service duct external to the modular unit provides access for maintenance. A steel walkway landing is manufactured and is attached to the module off site. It folds down on site and connects to the walkway of the next unit.

9.4.2 Concrete prison modules

Precast concrete modules are widely used for secure accommodation, such as prisons. The walls between cells and the roof are cast in one concrete pour using special moulds, which are designed to automate the de-moulding process. Window grilles and door frames are cast into the concrete walls, as shown in Figure 9.6. A seamless surface of floor, wall, and ceiling is achieved, which is important for both structural integrity and security. The manufacturing process used to produce the precast modular units follows a 6-day cycle to allow for concrete curing, etc.

A typical precast concrete module may contain two or four cells and weighs about 40 tonnes. Precast concrete prison cells are generally manufactured with an open base, and the ground floor module is placed on a prepared ground slab. The roof of the lower unit forms the floor of the upper unit. The external services between the cells are then connected from an accessible service area, which is described in more detail in Chapter 15.

A concrete walkway may be cast as part of the cell module. Figure 9.7 shows a module comprising a pair of cells being lifted into place. A rear chase runs along the exterior wall of a row of cells and provides for maintenance access to the mechanical systems without having to disrupt the prison's daily routines. The exterior wall is then insulated and clad to enclose the rear chase area. Figure 9.8 shows a completed group of units.

Figure 9.7 Window grilles cast in for security. (Courtesy of Precast Cellular Structures Ltd.)

Figure 9.6 Construction of HMP Rochester. (Courtesy of Britspace.)

Figure 9.8 Concrete cell units. (Courtesy of Oldcastle Precast.)

CASE STUDY 33: MODULAR SUPERMARKET DELIVERED TO ORKNEY

Installation of single-storey modules for the supermarket. (Courtesy of Yorkon.)

Shipping of modules to Orkney. (Courtesy of Yorkon.)

Supermarkets are increasingly constructed in fully modular form for speed of installation and to reduce commissioning and disruption during the construction process. This is particularly important where conventional forms of construction are more difficult, such as next to existing supermarkets, or in remote locations. In a project for Tesco in Kirkwall, Orkney, 26 purpose-built modules were delivered by road and shipped from Yorkon's factory in York to the port of Wick and then to the Orkneys.

The construction period for the 2500 m² supermarket was reduced to 3 weeks, and importantly, the existing supermarket on the site could be kept operational while the modules were installed and the building finished internally. The contractor for this project was Barr Construction.

The modules were 15 m long by 3.3 m wide by 3.6 m high internally and were manufactured as fully open-sided, except for the modules on the perimeter of the building, which had infill walls. Two modules placed longitudinally formed the 30 m wide supermarket. The 100 × 100 square

hollow section (SHS) corner posts of adjacent modules are clustered together to form encased columns at suitable locations within the store layout. Service routes were incorporated within the modules to suit the aisle positions in the supermarket.

Structurally, the 350 mm deep C section roof beams in the modules spanned 15 m, but the similar-sized floor beams were supported by concrete footings at 3 m spacing. Design floor loading was 4 kN/m², which is suitable for a wide range of applications. Stability was provided by separate vertical bracing in the perimeter walls of the modules, and wind forces were transferred by the connections between the modules.

Facilities in the supermarket also included a chiller section, office, storeroom, storage, and plant room, which were fully fitted out before delivery. The shipping of the modules to Orkney from Wick is shown above. Yorkon has installed more than 200 similar single-storey convenience stores, kiosks, and filling stations in the UK.

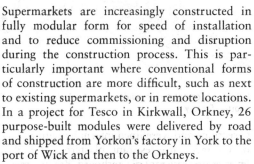

CASE STUDY 34: MILITARY ACCOMMODATION, NEAR SALISBURY

External view of military accommodation.

Lifting of completed module in Tata Living Solutions' Shotton factory.

The Ministry of Defence (MoD) commissioned Aspire Defence Ltd. to provide high-quality accommodation for military and civilian personnel near Salisbury Plain and Aldershot. Corus Living Solutions (CLS) was awarded the subcontract to design, manufacture, and install modular accommodation units. The initiative provides living and working accommodation, in a campus-style environment, for some 18,000 military and civilian personnel and includes 10,700 single living accommodation (SLA) units.

In 2005, CLS built its first accommodation block for junior ranks at Perham Down, Wiltshire, prior to the construction of 145 similar blocks over the next 6 years. The 3-storey block provides accommodation for 36 soldiers. The building was split into six flats, with each flat housing six people, together with a plant room located in the roof space. Each bedroom has an en suite shower room and use of a communal lounge.

The building comprises 51-room modules manufactured at CLS's semiautomated production line in Shotton, north Wales. The room modules were delivered to site as fully fitted out and serviced building blocks. The modules for this building were installed in just 5 days.

Each bedroom module is 3.38 m wide by 4.96 m long (internal dimensions). The wall panels were constructed from 100 × 55 × 1.8 mm

C sections and were lined internally with a layer of foil-faced fire-resistant plasterboard and a layer of fire-resistant plasterboard. Bracing members placed in the walls of the modules resist wind loading.

The decision to use modular construction rather than traditional methods was primarily influenced by the client's requirements for rapid speed of construction combined with the complex logistics of bringing construction personnel to secure MoD sites. Additional client requirements to minimise disruption on these actively used sites and to meet high standards of quality were more easily met using modular construction. Added benefits were economy in manufacture, as well as single-point procurement.

Externally, the module walls are sheathed with moisture-resistant plasterboard. The building is then clad with brickwork and rendered blockwork. The top of each module was protected during transportation and for the short time the module was left on site before the permanent roofing was installed.

This system of construction can be designed for buildings up to 5 storeys' height, depending on the wind load acting on the modules. Many other military accommodation projects have been procured using modular systems from a range of suppliers.

CAST STUDY 35: MILITARY HOUSING IN MIDDLESEX AND YORKSHIRE

View of 2- and 3-storey buildings from the courtyard. (Courtesy of Caledonian Modular.)

Modular construction has been extensively used for military accommodation in the SLAM and Aspire programmes because of its greatly reduced time on site and the small number of security-cleared personnel that are required in the off-site construction process. The redevelopment of Northwood HQ in Middlesex provided high-quality accommodation for 199 senior ranks and 279 junior ranks, together with communal and utility spaces.

Caledonian Modular manufactured and installed the 315 modules, and also designed and placed the roofing and wall cladding, and installed all the mechanical and electrical services. This was because a single source of procurement was important to the project manager and client, Carillion. The modules were 3.6 m wide and ranged from 12.3 to 17.2 m, including a central corridor. The accommodation was arranged in single- and double-module formats, and the building heights ranged from 2 to 5 storeys.

The overall construction period for this complex logistical project was 18 months, which represented a significant savings on more traditional construction. It achieved BREEAM "Excellent"

as the sustainability standard, mainly because of the high level of thermal insulation and airtightness provided. The modules also incorporated a mechanical ventilation and heat recovery system with integral ducting. Cladding was in the form of insulated render with featured timber panels. The modules were manufactured to achieve blast resilience by an additional external layer in their manufacture.

Other projects were completed by Caledonian Modular at RAF Wittering and Leconfield, Marne Barracks at Catterick, and an estimated 10,000 modules have been built and installed as part of the SLAM initiative over the last 8 years. The project at RAF Leconfield in east Yorkshire was also managed by Carillion and also included a 2-storey medical facility. The monopitch roof was manufactured as part of the 3.6 m wide modules, which were delivered at their maximum height of 4.1 m for road access. The ward space was built using open-sided modules, but the specialist rooms were built using single or double modules. It was procured under the maximum price target cost (MPTC) contract with the Ministry of Defence and included all the required medical fitments.

CASE STUDY 36: MODULAR PRISONS IN NOTTINGHAM AND LIVERPOOL

Precast concrete modules during installation. (Courtesy of Pre-cast Cellular Structures Ltd.)

Concrete modules are often used in secure accommodations, prisons, and other high-security applications, such as Ministry of Defence buildings, as they are extremely resistant to damage. Composite Ltd., based in Southampton, set up Pre-cast Cellular Structures Ltd. (PCSL) to design, manufacture, and install concrete modules for the prisons sector. PCSL manufactures at two sites in the UK. Approximately 2000 concrete modules have been manufactured by PCSL to date, one of the largest being a 600-unit prison in Liverpool. Another project in Nottingham consisted of 340 units.

Prisons and secure accommodation are procured on a turnkey or design-and-build basis, which includes internal facilities and fixed security systems. A 38-week construction programme can be achieved for a 180-cell block. Furthermore, the installation of the brick cladding and other external features takes place off the critical path, as it can be carried out as the weathertight modular building is being finished internally.

Cells are often manufactured in multiples within one module to maximise efficiency. Toilet blocks, side rooms, and multifunctional rooms are also delivered as complete modular units. Core areas often use L- and T-shaped wall panels for stability during installation. Designs up to 5 storeys high can be easily achieved. Innovations also included under-floor heating by embedding pipes in the slab.

Installation of the modules, which can weigh up to 25 tonnes, is generally by a 150-tonne capacity crawler crane, which is strategically located on the project site. The roof structure is often fabricated in steelwork and is supported by the precast walls or by the modular precast units themselves.

REFERENCES

Brooker, O. and Hennessy, R. (2008). *Residential cellular concrete buildings*. A guide for the design and specification of concrete buildings using tunnel form, crosswall, or twinwall systems. CCIP-032. Concrete Centre, London.

WRAP. (2008). Woolwich single living accommodation modernisation (SLAM) regeneration. http://www.wrap.org.uk.

Chapter 10

Hybrid modular construction systems

Mixed-use buildings require provision of flexible space and a wider range of services than in single-use buildings. Open-plan space is generally required for commercial areas, whereas cellular space is for required toilets and specialist facilities. This means that an open or adaptable building technology using modular and other forms of construction is often required for these buildings. The range of structural systems that may be considered depends on the

- Relative requirements for open-plan and cellular space
- Repetitive nature of the cellular space
- Service requirements in the various uses
- Location of stairs, lifts, and vertical bracing
- Load-bearing capabilities of the modules
- Requirement for change of use in the future (retrofitting)

Modules (3D elements) may be combined with planar (2D) and skeletal (1D) elements to create more flexible and adaptable building forms, which are explored in this chapter.

10.1 MODULAR AND PANEL SYSTEMS

One form of hybrid construction is where modules are combined with planar load-bearing wall panels and floor cassettes, which is presented in SCI P-348 (Lawson, 2007). In this form of construction, the modules provide the cellular space that is often highly serviced, and the planar elements that provide the open-plan space. The modules are stacked vertically, and therefore support their own loads and the loads from the incoming floors. The floor cassettes should ideally occupy the same depth as the combined depth of the floor and ceiling of the adjacent modules.

Modules may be used for kitchens and bathrooms, which should be located so that they can be combined in a module and so that service routes that can be accessed for maintenance. The combined depth of the floor and ceiling of the modules may be around 450 mm. This type of construction was first used in the Lillie Road

project in Fulham, shown completed in Figure 1.14 and during construction in Figure 10.1. The load-bearing modular bathrooms are shown in the foreground and the X-braced light steel walls to the side.

Modular supplier Elements Europe has developed a system called Strucpod that uses load-bearing bathroom modules in combination with light steel framing, which is based on a similar concept. It has been used in an 8-storey student residential building; see case studies in this chapter.

A demonstration project by Tata Steel (European Commission, 2008) used long composite timber-steel joists built into floor cassettes, which spanned between steel posts located within the modules. This system was designed to create an urban terrace streetscape, as illustrated in Figures 10.2 and 10.3.

This hybrid technology of planar and modular construction has four main components:

- Bathroom, kitchen, and stair modules that provide the stability of the building assisted by the planar walls
- Long spanning floors that provide the flexible living space with freedom in internal partitioning
- Load-bearing planar cross-walls
- Non-load-bearing façade walls

In order to maximize the plan area of the building devoted to adaptable living space, one stair/lift module accessed the adjacent two apartments at each level. The stair modules are 2.6 m wide and 10.5 m long. The modules comprising the bathrooms and kitchens for two adjacent apartments are 4.3 m wide and incorporated a separating wall internally. The kitchens were open-sided, so that they formed an open-plan space with the living area. This was achieved by using intermediate square hollow section (SHS) posts in the 5.5 m long kitchen/bathroom modules. The floor of the module was 200 mm deep, and its ceiling was 150 mm deep, and the combined depth of the floor and the ceiling in the modules was 450 mm.

The floor depth of the cassettes was chosen to be equal to the combined floor and ceiling depth of the module,

Figure 10.1 Combined modules and panels in the Lillie Road, Fulham, project (shown completed in Figure 1.14).

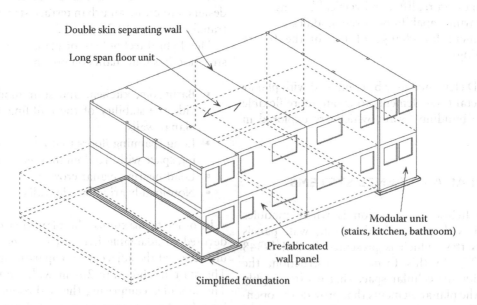

Figure 10.2 Urban terrace using hybrid modular and panel construction.

and so spans of up to 6 m between the module and the load-bearing cross-walls could be achieved. The living space created could therefore be adapted to the user requirements, as illustrated in Figure 10.4. In this layout, the ground floor was fitted out as a single-bedroom apartment, and the upper floors as two-bedroom apartments.

10.2 EXAMPLES OF PLAN FORMS USING HYBRID CONSTRUCTION

An example of a 3- or 4-storey residential building comprising adjacent apartments using mixed modular and planar elements is illustrated in Figure 10.5. The kitchen and bathroom modules are stacked vertically,

Figure 10.3 Demonstration building using hybrid modular and panel construction.

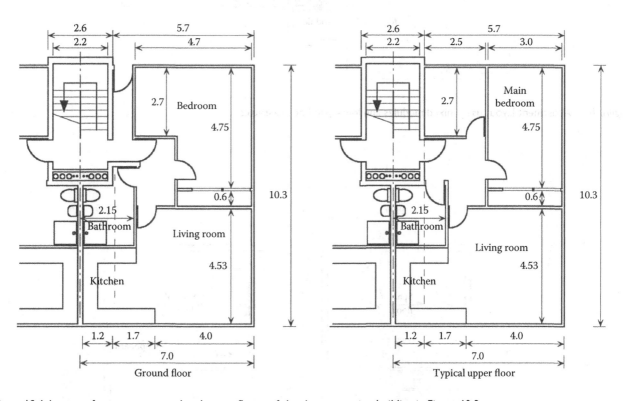

Figure 10.4 Layout of rooms on ground and upper floors of the demonstration building in Figure 10.3.

and floors span directly between the gable or separating walls, and are also supported by the module. In this example, the stairs are also installed as a separate module. Of the 54 m² floor area in each apartment, the modules comprise approximately 20% of this area. The double-leaf walls in modular construction provide the necessary acoustic separation and fire compartmentation functions between apartments.

In 2- or 3-storey housing, the kitchen and bedroom modules may be stacked vertically, as in Figure 10.6, and the side of the module forms part of the separating wall between terraced houses. In this example, the floors span 4.8 m between the cross-walls; the stairs are not made as part of a module and are located transverse in the building. A cross-section through a 3-storey house using this form of module is illustrated

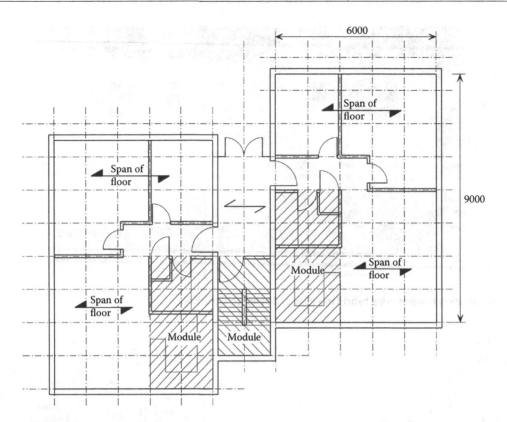

Figure 10.5 Apartment layout using mixed modules and long-span floor cassettes.

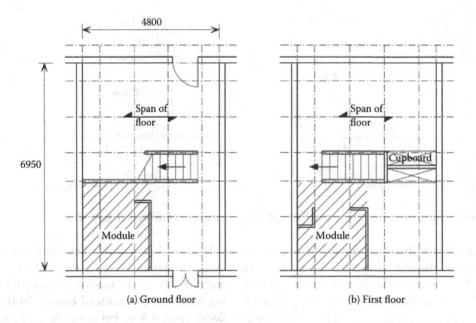

Figure 10.6 House layout using mixed modules and floor cassettes.

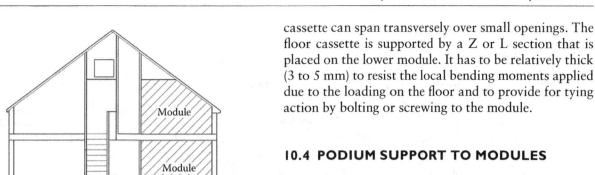

Figure 10.7 Cross section through house showing module positions.

in Figure 10.7. The upper module may be manufactured with a mansard roof.

10.3 DETAILS OF CONNECTIONS BETWEEN MODULES AND FLOOR CASSETTES

The details of a long-span floor cassette that is supported by the modules are illustrated in Figure 10.8. The floor joists are made as lattices of the exact depth to match the combined depth of the floor and ceiling of the module. The lattice is itself made from C sections of 70 or 100 mm width. The floor cassette is typically 350 mm deep, so that the overall floor depth is 450 mm, including the plasterboard below the joists and the acoustic layers above. The perimeter C section in the floor cassette is sufficiently stiff, so that the

cassette can span transversely over small openings. The floor cassette is supported by a Z or L section that is placed on the lower module. It has to be relatively thick (3 to 5 mm) to resist the local bending moments applied due to the loading on the floor and to provide for tying action by bolting or screwing to the module.

10.4 PODIUM SUPPORT TO MODULES

Modules may be supported on a steel or concrete podium structure that is generally located at the first or second floor level. In this way, the space below the podium level can be configured to suit its intended use. The supporting beams in the podium structure should align with the load-bearing walls of the modules above, as shown in Figure 10.9(a). In the case shown, the long-span cellular beams are supported on perimeter columns that align with the module width of 3 to 3.6 m.

An alternative arrangement in terms of use of space is shown in Figure 10.9(b). In this case, the columns align with twice the module width. For ground floor or basement car parking, the optimum column spacing is 7.5 m, which provides space for three car widths. This means that the modules may be 3.7 m wide externally (allowing for a 50 mm gap between them). For two car park spaces, the optimum column spacing is 5.4 m, which means that the modules are approximately 2.7 m wide. This is less efficient for housing, but is more suitable for student residences.

Deep cellular beams may be designed to span up to 16.5 m, which is the optimum for two lines of car parking and a central aisle without internal columns. However, the maximum number of levels of modules that may be supported in this long-span system is six or seven, as the beams will be up to 1 m deep and the slab will be a further 150 to 180 mm deep.

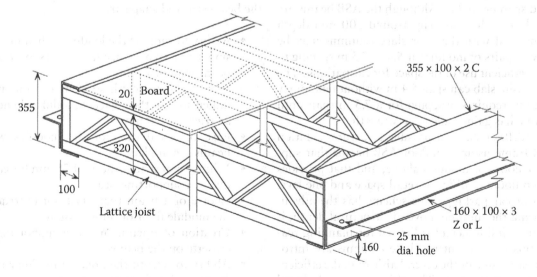

Figure 10.8 Details of long-span floor cassette supported by modules.

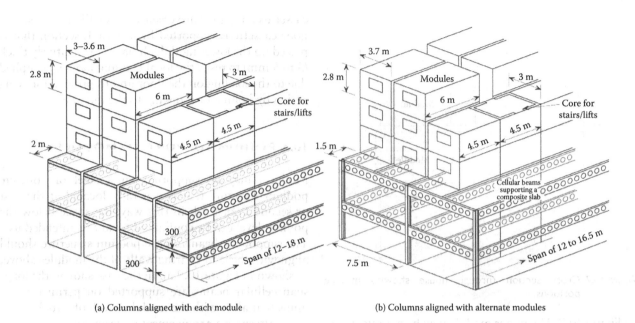

Figure 10.9 Podium structure composed of long-span cellular beams to support modules at the second floor level. (a) Columns aligned with each module. (b) Columns aligned with alternate modules.

The combined weight of the modules is applied to the supporting beams. For 6 floors of modules, the line load per beam can be up to 10 tonnes per m length when using light steel modules, and higher if the modules have concrete floors. The supporting beams should generally be designed as 'key elements' for structural integrity in accidental damage scenarios and also at the fire limit state. The building should be stabilised by concrete core or braced steel frame, whose position is optimised with respect to the use of the car park space below.

An alternative structural system used to support the modules may be in the form of slim floor beams together with a deep composite slab spanning between the beams. The slim floor beams (known as Asymmetric Slimflor® beams [ASB]) are integrated within the slab depth and span up to 8 m. Although the ASB beams are relatively heavy, they are only around 300 mm depth when combined with the floor slab. Columns may be aligned with pairs of modules at 5.4 to 7.5 m spacing to permit the efficient use of the space for car parking. The deep composite slab can span 5.4 m without temporary propping but requires propping for a 7.5 m span until the concrete has gained adequate strength.

Figure 10.10(c) shows a possible structural solution of a steel frame using slim floor ASB beams that supports four floors of modules above, and that provides one or two floors of office or retail space and one level of basement car parking. This scheme has the advantage of being the minimum overall structural depth, if the overall building height is limited for planning reasons. In this scheme, intermediate columns are introduced on either side of the central aisle to make efficient use of the car park space. The office space is designed

with a central line of columns, and a deep transfer beam spreads the load from the central column to the columns on either side of the aisle on the car parking level. The dimensions of the transfer beam with its regular castellations below the ASB steel section are indicated in Figure 10.10(b).

10.5 INTEGRATED STEEL FRAMES AND MODULES

In multistorey modular construction, in which the modules are supported by an independent steel frame, various types of beams may be used to support the modules at each floor level. The factors that control the choice of the beam size and shape are

- Ability to support the loading in bending and torsion from two adjacent modules over a typical beam span of 7.5 m
- Narrow beam width so that the combined wall dimensions of the adjacent modules do not exceed a target of 300 mm
- Ease of installation of the modules when the beams are in place
- Minimum bearing length of 75 mm for each module on a supporting beam
- Choice of a beam that must not protrude above the module floor or floor cassette
- Creation of open-plan space supporting a floor cassette on the beams
- Ability to retrofit the areas of modular units into an open-plan floor in the future

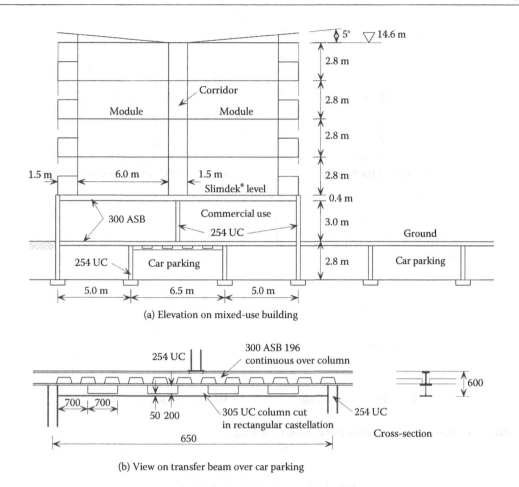

Figure 10.10 Podium structure composed of slim floor beams over an office and basement car park.

In this form of construction, stability is provided by the steel frame by bracing between the columns or around the lift core. Also, the steel frame may project outside the line of the modules to accommodate enclosed walkways or balconies. A variety of modular and open-plan spaces can be created on a particular floor, and the modular arrangement may be varied from floor to floor. In this way, a structural system is created that is adaptable to a range of uses.

The simplest structural system uses wide-flange beam sections to support two modules. A 254 mm steel universal column (UC) or H section is the sensible minimum beam width, and it can span up to 8 m when supporting two modules. In addition, modules can be manufactured with reentrant corners so that they fit around the columns in order to minimise the gap between the modules. In areas requiring open-plan space, the floor cassettes are designed to span between the beams, and so the modules and floor cassettes are interchangeable.

The beam design must take into account combined bending and torsional actions due to unbalanced loads transferred from the modules, and must not deform excessively due to loads applied during installation. Various types of beams may be used to minimise the

combined depth of the beam and module floor and ceiling, as shown in Figure 10.11. In this case, the bottom flange has to be narrow to fit between the modules, or alternatively, the ceiling of the modules has to be recessed. In this system, the combined floor and ceiling depth is around 500 mm.

The details of a light steel module with recessed corners and ceiling that can be integrated efficiently within a steel frame are illustrated in Figure 10.12. The light steel framework of the module may be minimised (for example, by reducing the steel thickness to 1 mm) because it is designed to be supported by beams on every floor. However, it is still required to be sufficiently stiff for transport and installation.

10.6 EXAMPLE OF MIXED USE OF MODULES AND STRUCTURAL FRAMES IN A RESIDENTIAL BUILDING

As an example of how modules may be combined with structural frames, the residential building layout in Figure 10.13 shows the serviced kitchen and bathroom modules arranged along the spine of the building.

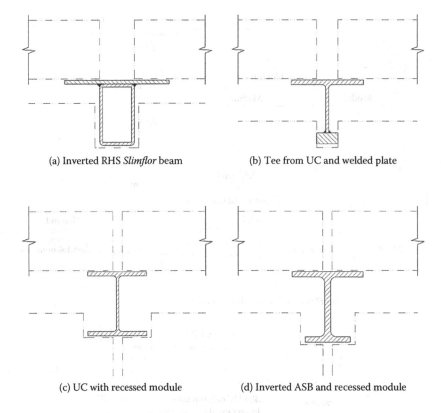

(a) Inverted RHS *Slimflor* beam

(b) Tee from UC and welded plate

(c) UC with recessed module

(d) Inverted ASB and recessed module

Figure 10.11 Beams providing support to the modules on their top flange.

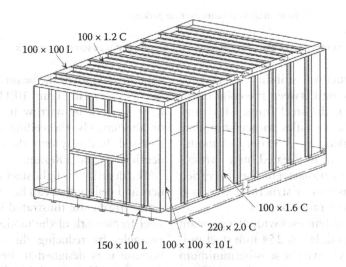

100 × 1.2 C

100 × 100 L

150 × 100 L   100 × 100 × 10 L

220 × 2.0 C

100 × 1.6 C

Figure 10.12 Module with recessed corners and edges supported by a steel frame.

The open-plan space is created by a composite floor system that spans between beams at 7.5 m spacing. Light steel partitions form the rooms in the living space and corridors. In this building form, the modular bathrooms and kitchens are also supported by the beams, and their position on the plan dictates the layout of the apartments. The modules can be serviced from the corridors.

In this example, the kitchen/bathroom modules are 2.1 m wide and 7.5 m long and are supported on slim floor beams, which also support the 300 mm deep floor slab. The beams may be in the form of rectangular hollow sections (RHSs) with a welded bottom plate or ASB slim floor that are also up to 300 mm deep. These beams support either a deep composite slab or precast concrete units with a clear span of up to 7.2 m allowing for the width of the beams.

Square hollow sections (SHSs) of 150 or 200 mm width are sized so they fit in the walls and in the recessed corners of the modules (see Figure 10.12). They are placed on a 7.5 × 6.7 m and 7.5 × 4.8 m grid

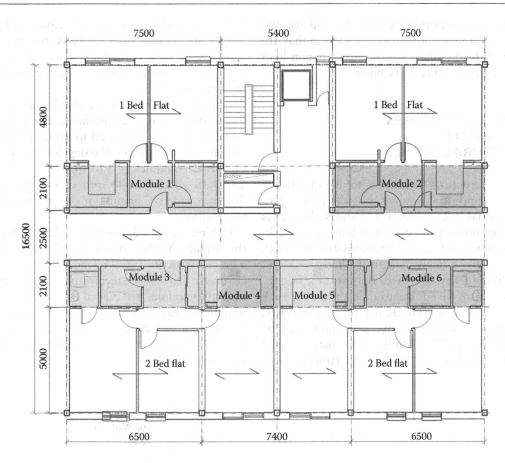

Figure 10.13 Mixed use of a structural frame and modular units for the serviced areas in a residential building.

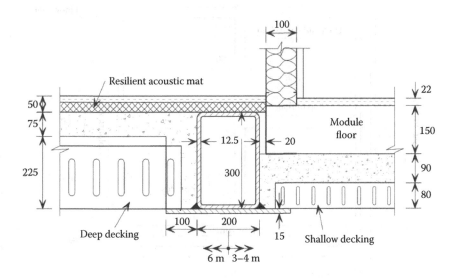

Figure 10.14 Module placed in a recessed floor and supported by a RHS slim floor beam.

to optimise the space if basement or ground floor car parking is required. The building depth of 16.5 m provides a 6.7 m wide access aisle, and the column spacing provides three car park spaces between the columns. A total of 15 car park places is provided in this layout of four apartments per floor, and so a single level of parking is suitable for a four-storey building.

Figure 10.14 shows the details of the alternative use of 300 × 200 RHS beams that support a 300 mm deep composite slab in the general floor areas and a 170 mm deep composite slab in the modular areas. The floor of the module is designed to be at the same level as the finished floor of the open-plan space, including the built-up layers for acoustic insulation. In this case, the 170 mm deep slab provides support to the modules and also the required fire resistance and acoustic insulation that are independent of the modules. The modules are non-load bearing when supported at every floor.

10.7 GROUPS OF MODULES SUPPORTED BY STRUCTURAL FRAMES

An alternative structural solution is to design a structural frame to support a group of modules rather than single modules. For a 2 × 2 group of modules, the beams are designed to support the weight of four modules. In this way, the depth and size of the beams are increased, but conversely, these beams occur at alternate floors, and so the floor depth is increased only at these floors.

The architectural design has to reflect the difference in the floor levels in the façade details, in the stairs, etc. For example, the supporting steel frame can be expressed externally. This configuration may be used efficiently for duplex or 2-storey apartments with balconies at the beam level. Because of the insulated nature of the individual modules, the steel beams and columns do not contribute significantly to thermal bridging, and projecting beams can be used to support the balconies directly.

Figure 10.15 shows a possible layout of a structural frame supporting four modules. The combined beam and floor zone is increased to around 750 mm, but at the intermediate floors, the depth of the floor and ceiling of the modules is 300 mm. A combined column and wall width of 600 mm should be allowed for a 300 mm wide UC or tubular column at alternate module positions.

In New York, a 32-storey residential building called the Atlantic Yards is in construction, which consists of a braced structural steel frame that supports the modular units. The braced steel frame is installed at the same time the modules are placed and is fitted between the modules at each floor level. The columns, beams, and V-bracing are external to the modules, and so the structure is part of the external architecture of this highrise building.

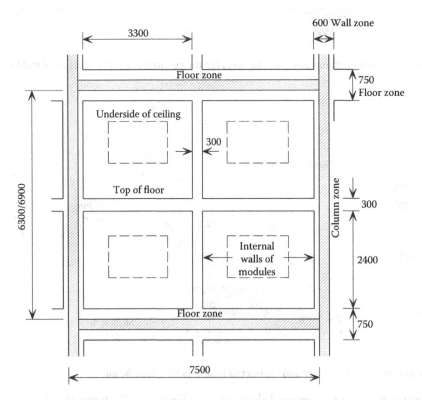

Figure 10.15 Dimensions for a group of four modules supported by a primary steel frame.

CASE STUDY 37: SOCIAL HOUSING ABOVE OFFICES, EAST LONDON

View of mixed-use building from Commercial Road, east London.

Modules with the braced steel access cores at one end. (Courtesy of Rollalong.)

This £5 million social housing project on London's Commercial Road for Tower Hamlets Community Housing (THCH) was constructed using modules placed on a reinforced concrete podium. The L-shaped residential building, called Painter House, was built using 76 modules that cascade from 5 to 2 storeys high on the first floor podium. The THCH occupies the office space at ground floor. A steel frame at the corner of the building provides the access core and walkways to the upper levels.

A total of 24 one-bedroom flats each comprise two 3.6 m wide by 7.7 m long modules. The two bedroom flats comprise two 3.6 m wide by 10.6 m long modules of approximately 75 m², which provides two spacious double bedrooms. Both types have separate kitchen, bathroom, and spacious living area with access onto an integral balcony built as part of the modules. The modules are accessed by an external walkway on the courtyard side of the building.

The project started on site in mid-2005 with demolition of the existing building. The main contractor was the Hill Partnership. The 76 modules were installed over a 3-week period at times to suit the traffic on London's busy Commercial Road. It was completed in August 2006, which represented a saving of 6 months on conventional site-intensive construction.

The light steel modules were designed, manufactured, and fully fitted out by Rollalong. Modules were delivered "just in time" to site at a rate of eight per day, and were each delivered with a protective shroud that remained in place. The rain screen façade comprised Trespa lightweight panels at the upper levels and brickwork at the lower levels. The monopitch roof was clad in Kingspan composite panels with a secret gutter and were fixed to the top floor module. These panels and finishes were installed from the scaffolding, which was attached to the modules.

A steel-framed stair and lift tower at the end of the building was constructed at the same time as the installation of the modules, and the steel frame was extended to provide access to each floor at the rear courtyard. The stair access towers and ground floor were brick-clad.

A single vertical group of five modules was designed to be stable under wind loads. The modules were tied together by cruciform connectors at their corners in order to transfer wind forces, and also to provide alternative load paths in the event of accidental damage to the ground floor. The fitted-out modules each weighed approximately 8 tonnes, and the concrete slab at podium level was designed to support the loads from five modules above.

CASE STUDY 38: MIXED HOUSING AND HOTEL DEVELOPMENT, WEMBLEY

View of completed hotel, Wembley.

Installation of a module with a concrete base.

An innovative mixed private and affordable housing and hotel project was completed in May 2013 on Olympic Way, Wembley, in north London. A total of 158 apartments consisting of 68 one-bedroom, 71 two-bedroom, and 19 three-bedroom apartments were designed along the adjacent Fulton Road and in the main 20-storey oval-shaped block. The tenure mix is 83% private and 17% affordable. The 234-room 10-storey hotel occupies the main frontage to the building, and retail units are located below the hotel.

Architect HTA worked closely with the developer of the Wembley area, Quintain, and with the contractor and modular supplier, Donban, to create an exemplar of a mixed-use development in this important site next to Wembley stadium. The buildings were designed at the outset to use the Vision modular system. The modules in the hotel and medium-rise parts of the building were placed on a 1 m thick concrete podium, and the modules in the tower were placed around a 8.2 × 6.8 m concrete slip-formed core that was constructed in advance of the installation of the modules.

The Vision modules were manufactured up to 3.9 m wide and 12 m long and consisted of a concrete floor with a tubular steel framework. Two modules formed a one-bedroom apartment of 59 m² floor area with an integral glazed winter garden balcony, and three modules formed a two-bedroom apartment of 76 m² area. The module concrete floor was also extended to form the corridors and patio areas so that no site work was required to form each floor. Some modules had partially open sides to create larger spaces. On the hotel side, all the modules were built with a southerly oriented oriel window, which provided views of the stadium.

The cladding to the hotel and street-side apartments consists of horizontal terra-cotta tiles that were fixed to the face of the modules via secondary rails. In the façade walls, the window positions were offset on each floor. This is possible in modular construction because the side walls provide the load-bearing function. The oval-shaped tower was clad in vertically oriented glazed panels that had a mixture of 20, 40, and 60% light transmission.

Installation of the modules started in early 2012, and it took only 9 months for the 700 modules to be placed in three stages, starting first with the hotel and ending with the tower block from July to October 2012. Modules weighed up to 20 tonnes, and their delivery was carefully timed to minimise any interference with the use of Olympic Way. No installation took place during the Olympic fortnight. The estimated overall time saving for this project was 9 months relative to reinforced concrete, which was important for the opening of the hotel.

CASE STUDY 39: STUDENT RESIDENCE USING BATHROOM PODS, LONDON

View of curved façade from Whitechapel Road, east London. (Courtesy of Elements Europe.)

Element Europe's Strucpod system was used to construct an 8-storey student residence on Fieldgate Street, off the busy Whitechapel Road in central London. The building, called the Curve, was designed to create a seamless junction between two streets, and it proved to be more efficient to use structural bathroom pods in combination with light steel framing to form the various room shapes on the curved façade. The student residence part of the building was constructed on a 2-storey podium level that housed a metro supermarket and offices below. The light weight of the super-structure also made this podium level thinner than it would have been.

A total of 343 student rooms were created, which included kitchens and communal facilities. The Strucpod system uses the bathroom pods as load-bearing elements to support the floor joists so that the combined depth of the floor and ceiling of the modular pod is compatible with the finished depth of the adjacent rooms. It achieved flexibility in planning of the room layout for the curved façade.

In common with other similar student residence projects, speed of construction was important to the success of the system. The project started on site in July 2011 and was completed in March 2012, which represented a savings of 6 months on conventional site-intensive construction. The main contractor was MACE, and the architect was Axis. The package of the light steel framing and fitted-out bathrooms was £2.7 million out of a total project value of £20 million, which represented a considerable savings on site-intensive construction.

The light steel framing and bathroom modules were manufactured by Elements Europe in its factory in Shropshire, and so the interfaces and delivery were coordinated. The bathrooms were fully tested and finished before delivery to the site, and the services were accessed from the outside of the modules to avoid damage to the fitments inside. The 343 modules were installed over a 24-week period at times to suit the traffic on Whitechapel Road.

Other modular systems in the portfolio include Roompod, which is a room-sized modular system based on a light steel framework, Solopod, which is a modular bathroom used in concrete- and steel-framed buildings, and T-frame, which is a low- to medium-rise timber modular system.

REFERENCES

European Commission. (2008). *Promotion of steel in sustainable and adaptable buildings*. Report of ECSC demonstration project 7215-PP-058, report EUR 23201 EN.

Lawson, R.M. (2007). *Building design using modular construction*. Steel Construction Institute P348.

Tata Steel. *The Slimdek® manual*. www.tatasteelconstruction.com.

Chapter 11

Acoustic insulation in modular construction

Acoustic insulation is an important design requirement in residential buildings, hospitals, and schools. This chapter reviews the principles of acoustic performance in the context of the Building Regulations. The acoustic insulation of modular construction is improved by its double-layer wall and floor and ceiling construction. Details of acoustically conforming floor and walls are presented. The same details used to achieve the required acoustic insulation also contribute to effective fire resistance.

11.1 PRINCIPLES OF ACOUSTIC INSULATION

Sound levels and acoustic insulation values are expressed in decibels (dB), while pitch or frequency is expressed in Hertz (Hz) (BS EN ISO 717): in the case of sound levels, the decibel rating is a representation of the intensity or volume of the sound.

Acoustic insulation values are a measure of the amount by which sound that is transmitted from one area to another is reduced by the separating floor or wall. Acoustic insulation is measured at a number of different frequencies, usually 16 one-third octave bands from 100 to 3150 Hz, and is given as a single figure by comparing the actual sound reduction record over these frequencies with a series of reference curves. The acoustic insulation properties of wall and floor build-ups vary with frequency, and high-frequency sounds are normally attenuated (reduced) more than low-frequency sounds.

Sound insulation between rooms is achieved by applying the following principles in combination:

- Provision of mass of the separating elements
- Isolation of separate layers
- Sealing of joints and any gaps

The principles of mass and isolation of layers are shown in Figure 11.1. Sound transmission across a solid wall conforms to what is known as the mass law. This principle means that doubling of mass of a solid element will increase its acoustic insulation by approximately 5 dB. However, increasing the mass without the use of separate layers has a diminishing benefit in terms of acoustic insulation.

Sound absorption is a property in which the level and quality of noise within the space is controlled by reducing the buildup of noise through reverberation. When a room is separated from another room, sound can travel by two routes: directly through the separating structure, called direct transmission, and around the separating structure through adjacent building elements, called flanking sound transmission.

Direct sound transmission depends upon the properties of the separating wall or floor and can be estimated from laboratory measurements. Flanking transmission is more difficult to predict because it is influenced by the details of the junctions between the building elements. It is therefore important that good acoustic detailing is used at the junctions, which can be achieved more successfully in modular construction by its off-site manufacturing process.

11.2 ACOUSTIC REQUIREMENTS AND REGULATIONS

In the UK, acoustic insulation requirements in housing and residential buildings are given in Approved Document to Part E of the Building Regulations, which address as the required performance of floors and walls that separate one dwelling from another, or a dwelling from a communal space. The requirements must be met for all rooms for residential purposes, includes rooms in hotels, halls of residence, residential homes, etc. Testing standards are defined in BS EN ISO 140-4 and 140-7.

In the Approved Document to Part E, the minimum airborne sound reduction level, $D_{nT,w} + C_{tr}$, now takes into account a correction factor for low-frequency sound, C_{tr}, and is applied to the basic airborne sound reduction, $D_{nT,w}$. The limit on $D_{nT,w} + C_{tr}$ is a minimum value because it applies to the difference in sound level between the sound source and the receiving room. The guidance in the former and current approved documents is compared in Table 11.1. It is not possible to directly compare the levels of sound reduction in the

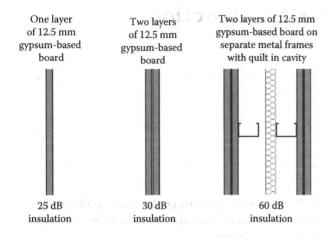

One layer of 12.5 mm gypsum-based board

25 dB insulation

Two layers of 12.5 mm gypsum-based board

30 dB insulation

Two layers of 12.5 mm gypsum-based board on separate metal frames with quilt in cavity

60 dB insulation

Figure 11.1 Sound insulation using mass and isolation of layers (sound reduction in dB).

Table 11.1 Minimum standards for acoustic insulation in the Building Regulations

	Separating walls		Separating floors		
	$D_{nT,w}$	$D_{nT,w} + C_{tr}$	$D_{nT,w}$	$D_{nT,w} + C_{tr}$	$L'_{nT,w}$
Former approved document E					
Average	>53 dB	n/a	>52 dB	n/a	<61 dB
Single value	>49 dB	n/a	>48 dB	n/a	<65 dB
Approved document E (2003)					
New-build dwellings (any test)		>45 dB		>45 dB	<61 dB
Conversions (any test)		>43 dB		>43 dB	<61 dB
Rooms for residential purposes (any test)		>43 dB		>45 dB	<61 dB

Source: The Stationery Office, Approved Document E, 2003 ed. (incorporating 2004 amendments), 2004.

Note: $D_{nT,w}$ = airborne round reduction, $L'_{nT,w}$ = impact sound transmittance, C_{tr} = correction factor for low-frequency sound (negative value).

regulations, because C_{tr} is a negative value, and C_{tr} can be in the range of −6 to −9 dB for lightweight floors and walls.

The impact sound transmission value $L'_{nT,w}$ is a permitted maximum value rather than a minimum value, as it applies to the sound that is transferred through the floor by a tapping machine. A double-layer floor and ceiling has superior impact sound reduction compared to a single-layer floor.

11.2.1 Demonstrating compliance

Approved Document E describes two methods of demonstrating compliance with the Building Regulations:

by precompletion testing (PCT) or by use of robust details (RDs). PCT is carried out on site, and the onus is on the builder to demonstrate compliance. PCT should be carried out when the rooms either side of the separating element are essentially complete, except for decoration. Buildings with rooms for residential purposes (i.e., hotels, student accommodation, etc.) are also subject to PCT.

Robust details only apply to houses and residential buildings, and a range of robust details (RDs) have been developed, which exceed the acoustic performance requirements specified in the regulations. Robust details are available for many forms of steel construction, including the double-layer walls and floors used in modular construction. Further information on acoustic insulation of various forms of steel construction is provided in SCI P372 (Way and Couchman, 2008).

11.2.2 Nonresidential buildings

For hospitals, acoustic requirements are specified in the Health Technical Memorandum (HTM) 08-01 (2008), which has replaced the former HTM 2045. For schools, Building Bulletin 93, *The Acoustic Design of Schools* (2004), produced by the Department for Education and Skills, should be adopted.

11.3 SEPARATING WALLS

11.3.1 Acoustic performance of light steel walls

In modular construction, each leaf of the wall is structurally and physically independent of the other, and so the overall performance can be approximated by simply adding together the sound insulation ratings of the two leafs.

The typical requirements for good acoustic insulation of separating walls in lightweight modular construction are

- No connections between the walls except through ties at floor levels
- A minimum weight of 22 kg/m² in each wall (i.e., two layers of 12.5 mm plasterboard or equivalent)
- Separation between the two plasterboard faces (200 mm is the recommended minimum separation)
- Good sealing of all joints that is achieved more reliably in prefabricated construction
- Mineral fibre quilt placed within both of the walls to reduce sound reflection between the C sections
- Sheathing boards placed on the outside of the module
- Optional resilient bars on the inside face of the C sections to support the plasterboard

Table 11.2 Acoustic performance of various light steel and modular walls

Construction	Performance
Twin light steel frames (quilt between frames)	$D_{nT,w} + C_{tr} = 45\text{–}56$ dB
Twin light steel frames for modular construction	$D_{nT,w} + C_{tr} = 48\text{–}56$ dB
Single light steel frame with resilient bars	$D_{nT,w} + C_{tr} = 47\text{–}51$ dB

Note: The acoustic performance will depend on several factors, including material specifications, workmanship, and detailing of joints.

The acoustic performance of typical light steel wall constructions is presented in Table 11.2. Special care should be taken around openings for service pipes and other penetrations. Electrical sockets penetrate the plasterboard layer and should be carefully insulated by quilt at their rear. Back-to-back electrical fittings should be avoided. Electrical wiring can be installed in preformed ducts in the factory, which facilitates commissioning on site and does not compromise acoustic performance.

11.3.2 Acoustic performance of concrete walls

In concrete construction, the mass of a wall or floor is primarily responsible for the sound reduction according to the acoustic mass law. Hence, precast concrete modules offer good airborne sound reductions, while impact sound is controlled by appropriate floor and ceiling finishes. The acoustic performance of typical

Table 11.3 Acoustic performance of concrete walls

Finish on one side	Structure	Finish on the other side	Airborne sound reduction
Paint finish to the concrete wall	150 mm solid precast concrete	Paint finish	45 dB
Two layers of 12.5 mm plasterboard supported by batten system with 70 mm Isowool in cavity	150 mm precast concrete	12.5 mm plasterboard on 38 × 25 mm battens	51 dB
2 mm plaster skim	180 mm in situ concrete	2 mm plaster skim	47 dB

Source: Brooker, O., and Hennessy, R., *Residential Cellular Concrete Buildings: A Guide for the Design and Specification of Concrete Buildings Using Tunnel Form, Crosswall, or Twinwall Systems*, CCIP-032, Concrete Centre, London, UK, 2008.

precast concrete walls is shown in Table 11.3, including the comparative data for a typical in situ concrete wall.

11.4 SEPARATING FLOORS

11.4.1 Acoustic performance of light steel floors

For a separating floor construction, both airborne and impact sound transmission should be addressed. High levels of acoustic insulation are achieved in lightweight floors by using a similar approach to that described for walls.

Airborne sound insulation in lightweight floors in modular construction is achieved by

- Structural separation between the floor and ceiling
- Appropriate mass in each layer due to the plasterboard ceiling, floor boarding, and the board on the roof of the module
- Sound-absorbent quilt between the C sections
- Minimising flanking transmission at floor-wall junctions

Impact sound transmission in lightweight floors is reduced by:

- Using a resilient layer with the correct dynamic stiffness under loading
- Isolating the floating floor surface from the surrounding structure at the floor edges

The acoustic insulation provided by a series of typical light steel floor constructions is presented in Table 11.4. A resilient layer beneath the floor finish reduces both airborne and impact sound transmission. Generally, mineral fibre of 70 to 100 kg/m^3 density provides sufficient stiffness to prevent local deflections, but is soft enough to function as a vibration isolator. Mineral wool placed within the depth of the floor and ceiling

Table 11.4 Acoustic performance of light steel floors and modular construction

Form of construction	Performance
Light steel joists with boards	Airborne: $D_{nT,w} + C_{tr}$ = 47–54 dB Impact: L'_{ntw} = 44–58 dB
Light steel joist and ceiling in modular construction	Airborne: $D_{nT,w} + C_{tr}$ = 48–55 dB Impact: L'_{ntw} = 50–60 dB

Note: Performance will depend on several factors, including exact material specifications, workmanship, and detailing of joints.

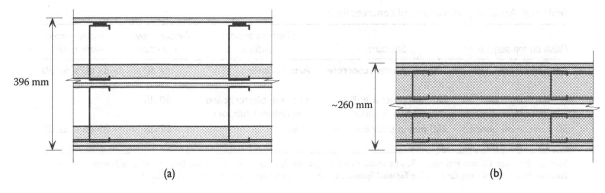

Figure 11.2 Details of modular construction used in the acoustic tests in Table 11.5. (a) Separating floor construction. (b) Separating wall construction.

helps to absorb sound in the cavity between the C sections. Also, increasing the mass of the top (floating) layer makes a significant improvement in the airborne sound insulation. Thicker layers of fire-resistant plasterboard have higher mass than ordinary plasterboard, thus reducing sound transmission.

Flanking transmissions can add 3 to 7 dB to the sound transfer measured in buildings in comparison to those tested acoustically in the laboratory. To reduce flanking transmission, it is important to prevent the floor boarding from touching the wall studs by including a resilient strip between the wall and floor boarding. Furthermore, in modular construction, the mineral wool insulation and sheathing boards help to reduce flanking transmission.

A series of acoustic tests were carried out in the completed modular rooms of the Paragon project in west London, which is shown in Figure 6.15. The form of light steel construction is illustrated in Figure 11.2, and the results are summarised in Table 11.5. The details are as follows.

Separating floor and ceiling (refer to Figure 11.2(a)):

- 18 mm flooring-grade plywood
- 3 mm isolation strip
- 180 mm deep-perimeter hot-rolled steel channel section with 165 mm light steel joists
- 80 mm of fibre glass insulation between the joists
- 150 mm deep-perimeter hot-rolled steel channel section supporting 138 mm steel joists boarded with 9 mm OSB board underneath

Table 11.5 Typical acoustic performance data for modular construction using a light-steel framework

Element	Sound reduction	Measured for modular construction	Regulations part E
Floor	Airborne	48 dB	$D_{nT,w} + C_{tr} \geq 45$ dB
Floor	Impact	54 dB	$L'_{nT,w} \leq 62$ dB
Wall	Airborne	47 dB	$D_{nT,w} + C_{tr} \geq 45$ dB

Source: Way, A.G.W., and Couchman, G.H., *Acoustic Detailing for Steel Construction*, Steel Construction Institute P372, 2008.

Separating wall (refer to Figure 11.2(b)):

- 1 layer of 12.5 mm fire-resistant plasterboard
- 1 layer of 12.5 mm standard plasterboard
- 80 mm light steel stud wall with 80 mm of glass fibre insulation between the C sections
- 4 mm sheathing grade plywood
- Cavity (30 to 50 mm wide)

11.4.2 Acoustic performance of concrete floors

The level of acoustic insulation provided by concrete floor slabs is presented in Table 11.6. In particular, the impact sound transmission is low even if the airborne sound insulation is moderate without a carpet or resilient layer.

Table 11.6 Acoustic performance of concrete floors

Finish on top surface	Structure	Finish on bottom surface	Airborne sound reduction	Impact sound transmission
Bonded carpet	200 mm precast concrete slab	Artex plaster	47 dB	34 dB ≤ 62 dB
65 mm screed on resilient layer	200 mm precast hollow-core concrete slab	12.5 mm plasterboard on channel support	50 dB	
50 mm screed bonded and 6 mm carpet	250 mm in situ concrete slab	None	57 dB	39 dB ≤ 62 dB

Source: Brooker, O., and Hennessy, R., *Residential Cellular Concrete Buildings: A Guide for the Design and Specification of Concrete Buildings Using Tunnel Form, Crosswall, or Twinwall Systems,* CCIP-032, Concrete Centre, London, UK, 2008.

REFERENCES

Brooker, O., and Hennessy, R. (2008). *Residential cellular concrete buildings: A guide for the design and specification of concrete buildings using tunnel form, crosswall or twinwall systems.* CCIP-032. Concrete Centre, London, UK.

British Standards Institution (BSI). (1997). *Acoustics. Rating of sound.* BS EN ISO 717.

British Standards Institution (BSI). (1998a). *Acoustics. Measurement of sound insulation in buildings and of building elements. Field measurements of airborne sound insulation between rooms.* BS EN ISO 140-4.

British Standards Institution (BSI). (1998b). *Acoustics. Measurement of sound insulation in buildings and of building elements. Field measurements of impact sound insulation of floors.* BS EN ISO 140-7.

Department for Education and Skills. (2004). *The acoustic design of schools.* Building Bulletin 93. The Stationery Office.

Department of Health. (2008). *Acoustics.* Health Technical Memorandum (HTM) 08-01. The Stationery Office.

The Stationery Office. (2003). *Building regulations (England and Wales) 2000. Part E. Resistance to the passage of sound.*

The Stationery Office. (2004). Approved document E. 2003 ed. (incorporating 2004 amendments).

Way, A.G.W., and Couchman, G.H. (2008). *Acoustic detailing for steel construction.* Steel Construction Institute P372.

Chapter 12

Structural design of light steel modules

Modules manufactured in light steel framing are usually designed to a standard specification for a particular project. For low- to medium-rise buildings, the structural design of the modules depends on the loading applied to the most highly stressed module at the base of the building.

For tall buildings, it is possible to increase the thickness of the steel or to reduce the spacing of the C sections at the lower levels. However, the manufacture of the modules is standardised over a number of levels (typically over four to six floors). This chapter presents the structural design of steel-framed modules in accordance with national standards and Eurocodes.

12.1 LOADING AND LOAD COMBINATIONS

The relevant national standards for the design of steel structures in the UK were BS 5950-1 and 5950-5 (BSI, 1997, 2000), but British Standards are now replaced by Eurocodes, for which BS EN 1993-1-1 and 1993-1-3: Eurocode 3 (BSI, 2004, 2006) are the comparable standards. The partial factors used to combine the effects of various loads are defined in Table 12.1. Imposed loads are variable loads due to the occupancy, whereas dead loads are permanent. Wind loads may be determined according to BS 6399-2 (BSI, 1997) or EN 1991-1-4: Eurocode 1: *Actions on Structures—General Actions: Part 1-4: Wind* (BSI, 2005) and its UK national annex. Wind loads are transient and conform to a 1 in 50-year recurrence wind speed.

Various load combinations are required by Eurocodes, depending on whether imposed load or wind is the dominant action. Generally, design to Eurocodes leads to lower factored loads when the design is controlled by vertical loading, but to higher factored loads when design is controlled by wind loading. This is because of the load factor of 1.5 for wind loading in BS EN 1991-1-4 in comparison to a factor of 1.4 in BS 5950. Also, the combination of vertical and horizontal loading is more severe in design to Eurocodes than to BS 5950.

The standard for imposed loading is BS 6399-1 (BSI, 1996), which is replaced by BS EN 1991-1-1: Eurocode 1: *Actions on Structures: Part 1-1: General Actions— Densities, Self-Weight, Imposed Loads for Buildings* (BSI, 2005), which are similar. Both standards permit the use of an imposed load reduction factor as a function of the building height. This takes account of the probability that all floors will not be loaded to their full design load. For a building of 5 or more storeys, the reduction in imposed loading acting together on all floors can be up to 40% when designing the vertical elements of the structure. This reduction does not apply to the self-weight and other permanent loads.

The design conditions that are considered in the structural design of light steel modules are therefore:

- Maximum vertical load when all modules are fully loaded, taking account of the self-weight of the modules, imposed loads, cladding, and roof loads.
- Maximum wind loads in combination with the self-weight of the modules, cladding, and roof, which influences the uplift on the foundations.
- Combined wind and vertical load with reduced load factors (given in Table 12.1) may lead to higher compression forces in the walls, or on the corner posts, than for the maximum vertical load case.
- Accidental load case where support to one module is notionally removed. The remaining group of modules must remain stable even under this extreme event, which is known as structural integrity or robustness. Overall stability in this condition is provided by the tying action between the modules.

In steel-framed modular systems, the means of resisting these loads depends on direct load transfer through the walls or occurs indirectly by edge beams, and then to corner posts. The following sections review the structural design of modular buildings using light steel construction, taking account of these load conditions.

12.2 FORMS OF CONSTRUCTION

The thin steel elements used in the floors and walls are generally of a C shape and are produced by cold rolling

Table 12.1 Load factors and load combinations to BS 5950 and Eurocode 3

Code	Load combinations	Loading		
		Imposed, IL	Dead, DL	Wind, WL
BS 5950	IL + DL	1.6	1.4	
	IL + DL + WL	1.2	1.2	1.2
	WL + DL	—	1.4 or 0.9	1.4
Eurocode 3	IL + DL	1.5	1.35	
	IL + DL + WL	1.5	1.05	1.05
	IL + DL + WL	or 1.05	1.05	1.5
	WL + DL	—	1.5 or 1.0	1.5

of a galvanised steel strip that is manufactured to BS EN 10346 (BSI, 2004b). These C sections are placed singly or in pairs at 600 mm centres in the walls and at 400 mm centres in the floors. Steel thicknesses of 1.5 to 2 mm are generally used in light steel framing and modular construction, although heavier sections can be used for the more highly loaded modules in buildings of 12 or more storeys' height.

12.2.1 Continuous longitudinal support to modules

In light steel modules, direct load transfer through the longitudinal sides of the modules can be achieved in various ways. The simplest way is to support the wall of the upper module directly on the wall below, as in Figure 12.1(a). The gap between the floor and ceiling is variable. Alternatively, ladder trusses can be built as part of the floor cassette and transfer the load from the walls above and below, as in Figure 12.1(b). In some systems, the edge of the floor cassette supports the wall, and is formed from a thicker C section so that it can transfer compression from the wall across its depth.

However, this system is limited to about 4 storeys' height because of the compressibility of the C sections in the floor cassette.

12.2.2 Corner support with edge beams

Corner-supported modules use longitudinal edge beams at floor and ceiling levels that span between the corner posts. The edge beams may be in the form of hot-rolled steel parallel flange channel (PFC) sections or various forms of cold-formed steel sections, which can be specially rolled for the particular type of module.

The use of PFC edge beams connected to square hollow section (SHS) posts is illustrated in Figure 12.2. In this case, the edge beam supporting the floor is 300 mm deep, and the edge beam at the ceiling is 200 mm deep, which leads to an overall floor depth of 600 mm, allowing for gaps. The longitudinal view of a module using the same PFC sections is illustrated in Figure 12.3. However, for modules longer than 7.5 m, it would be necessary to use deeper edge beams or to include intermediate posts to reduce the beam span and hence its size. The sides of the module may be infilled by light steel walls, or alternatively, the module may be delivered as open-sided to create open-plan space.

Cold-formed steel sections of 300 to 400 mm depth and 3 to 4 mm thickness may also be used for the floor and ceiling edge beams, but this leads to a relatively deep combined floor and ceiling depth (750 to 900 mm typically). The connections of the beams to the posts may be made by fin plates that are welded to the posts and bolted to the beams. These connections to deeper edge beams possess some bending resistance and can be used to provide stability to open-sided modules in low-rise buildings.

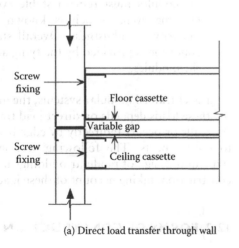

(a) Direct load transfer through wall

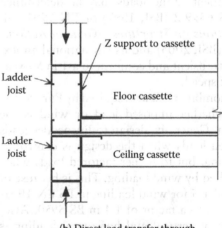

(b) Direct load transfer through ladder trusses in floor and ceiling cassettes

Figure 12.1 Direct load transfer through longitudinal edges of light steel modules.

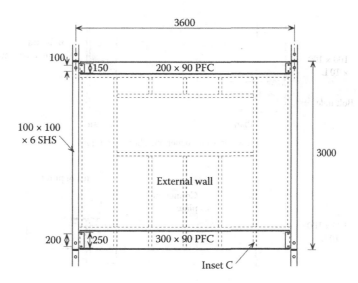

Figure 12.2 End view of corner-supported module using PFC edge beams.

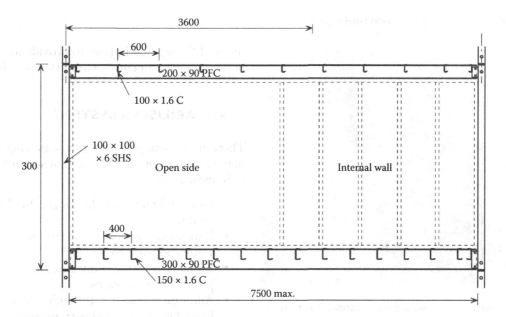

Figure 12.3 Longitudinal view of corner-supported module with a partially open side.

12.3 CONNECTION BETWEEN MODULES

For the corner connections between modules, two forms of corner posts may be used:

- Angle sections or other open cross sections
- Square hollow sections (SHSs)

Angle sections are relatively simple in that they may be introduced into the recessed corner of the module. The angle may be a cold-formed (i.e., bent plate of 3 to 6 mm thickness) or a hot-rolled section, typically a 100 × 100 × 10 mm thick angle. The modules may be

connected at their base and top and linked together by connector plates and single bolts, as in Figure 12.4(a). Alternatively, the angles can be connected by a side plate, as in Figure 12.4(b). In this case, a nut is welded to the face of the angle to allow the connection to be made from one side, as shown in Figure 12.5.

The load capacity of a steel angle is dependent on its size and whether it is stabilised by its connection to the walls of the module. For a partially open-sided module with a short length of wall next to the corner post, an angle section is relatively unstable in compression and is not recommended for buildings of more than 3 storeys' height.

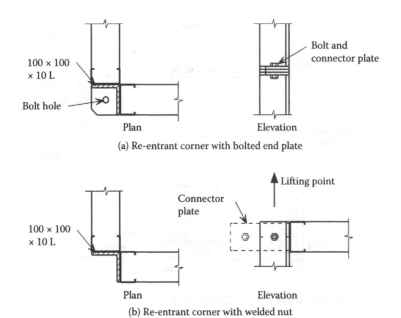

(a) Re-entrant corner with bolted end plate

(b) Re-entrant corner with welded nut

Figure 12.4 Corner posts using hot-rolled steel angles.

Figure 12.5 Corner angle with welded nut to connect the tie plate.

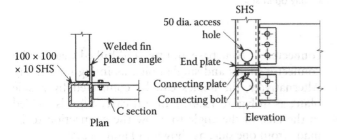

Figure 12.6 Corner post using SHS hollow sections.

SHS corner posts provide the highest compressive resistance and may be used for fully open-sided modules. Figure 12.6 shows a fin plate welded to the SHS to which the edge beams are bolted. Access holes of 50 mm minimum diameter in the SHS allow bolts to be inserted through end plates to provide for vertical and horizontal attachments between the modules.

12.4 STABILISING SYSTEMS

There are various generic forms of bracing systems that may be used to provide the overall stability of modular buildings, as follows:

- X- or K-bracing in the longitudinal walls of the modules
- Diaphragm action of sheathing boards to the walls with suitable fixings
- Moment-resisting connections between the corner posts and edge beams
- Additional stability through a concrete core or braced steelwork that is transferred by horizontal bracing in the corridors

12.4.1 X- or K-bracing

X-bracing in the form of cross-flats can be installed in the modules as part of the manufacturing operation, and may be used in the closed longitudinal sides of the modules. For X-bracing, the cross-flats are designed to resist only tension.

An alternative form of integral K-bracing may also be used adjacent to windows and doors where space is limited. The K-bracing members are C sections that are manufactured as part of the wall, and are designed to resist tension and compression.

Horizontal shear forces of the order of 20 kN can be resisted by X-bracing, and so this system is useful

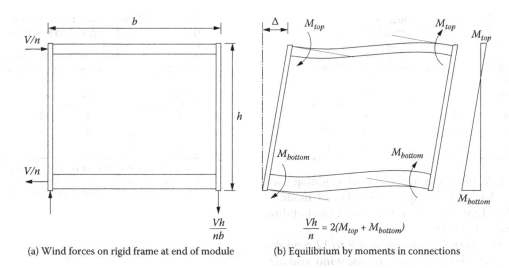

(a) Wind forces on rigid frame at end of module   (b) Equilibrium by moments in connections

Figure 12.7 Stability of low-rise building by moment-resisting connections between the corner posts and edge beams.

for medium-rise buildings (6 to 8 storeys). K-bracing resists much smaller shear forces of the order of 5 kN. It follows that K-braced panels placed on either side of a window opening at the end of the module can resist a horizontal shear force of about 10 kN.

12.4.2 Diaphragm action

Diaphragm action refers to the shear resistance of sheathing boards that are fixed to the light steel framework, and suitable sheathing boards, such as cement particle board (CPB), moisture-resisting plywood (WPB), orientated strand board (OSB), and moisture-resistant plasterboards (type H to BS EN 520). Guidance on diaphragm action is given by Lawson et al. (2005).

Diaphragm action of unperforated longitudinal side walls of a module leads to higher in-plane shear resistances than the equivalent X-braced walls. However, for effective shear resistance, fixings in the form of nails or screws should be installed at not more than 300 mm spacing on all sides of the boards. Tests on a 2.4 m² wall panel using the above type of boards show that CPB can resist a horizontal shear load of approximately 10 kN (or 4 kN/m wall length), which is controlled by a deflection limit of module height/500 (or approximately 5 mm) in the test. A comparable figure for OSB is 3 kN/m wall length. External sheathing boards also provide weathertightness in the temporary condition.

12.4.3 Moment-resisting connections

Moment-resisting connections may be in the form of end plate or deep fin-plated connections between hot-rolled steel posts (normally SHSs) and the edge beams (normally PFC sections), as illustrated in Figure 12.3.

Modules that are designed to be open-sided are mainly used for buildings up to 3 storeys high, unless some other stabilising system is provided. Relatively low moments can be transferred through the beam to post connections unless the fin plate is long enough to incorporate four bolts at as wide a spacing as is practical (typically 250 mm).

The rigidity of the longitudinal sides of the modules with moment-resisting connections can be improved by X-bracing, or by the use of intermediate posts. Connections to beams in the corridor zone can also assist in providing additional stiffness. The structural action of the end frame of an open-sided module subject to horizontal loading is illustrated in Figure 12.7. This can be achieved by a welded end frame using SHS sections. The moments in the connections are dependent on the number of modules, n, that resist the horizontal load and the height, h, of the modules.

12.4.4 Typical bracing requirements

Consider the stability of a group of four-sided modules used in a 13.5 m deep building of 4, 6, or 8 storeys' height and consisting notionally of 6 m long by 3.6 m wide modules placed on either side of a 1.5 m wide corridor. The group of modules is subject to an unfactored wind pressure in the range of 0.8 to 1.4 kN/m², which depends on the location and height of the building. Wind loading is considered first acting on the end gable, and then on the front and rear faces of the building. It is assumed that a roof of 30° slope spans from the front to the back of the building.

The minimum number of modules side by side in the group is calculated from a permissible shear load of 4 kN/m length of the walls on all four sides, based on

Table 12.2 Minimum number of modules placed horizontally in a group to resist wind forces on the end gable

Number of storeys	Characteristic wind pressure (kN/m²)			
	0.8	1.0	1.2	1.4
4	6	7	8	9
6	8	9	10	12
8	10	12	14	16

Table 12.3 Minimum number of modules in the building depth required to resist wind forces on the facade

Number of storeys	Characteristic wind pressure (kN/m²)			
	0.8	1.0	1.2	1.4
4	1	1	2	2
6	2	2	X	X
8	2	X	X	X

1 = one line of modules can resist the wind forces, 2 = two modules required on either side of a corridor, X = additional stabilising system required.

the normal deflection limit for brickwork cladding used on the lower levels. The required number of modules is presented in Table 12.2 as a function of the building height and wind loading.

This table shows that a minimum of 8 to 12 modules should be placed side by side to share the wind load acting on the end gable of a 6-storey building, assuming that the modules resist equal loads. It is possible to relax the horizontal deflection limit for lightweight cladding materials, and therefore to increase the permitted height by 1 storey typically for each case in Table 12.2.

Considering wind acting on the front or rear façade of the building, the sheathed longitudinal wall of a 6 m long module can resist a shear force of approximately 24 kN. Based on this permitted shear force, the maximum height of a building consisting of one or two modules in its depth is presented in Table 12.3. This shows that the maximum building height is typically 6 to 8 storeys. For the cases identified by X, wind loads should be transferred horizontally to cores or braced walls in order to provide overall stability of the building. This simple analysis shows that a 6-storey building in a corridor-style layout should comprise a group of at least 2 × 9 modules per floor or wing for wind loads up to 1 kN/m².

12.4.5 Horizontal bracing

Additional vertical bracing may be incorporated in a separate structural frame around stairs and lifts or in the end gables. In this case, horizontal bracing is required to transfer forces to these points, and this bracing can be incorporated in the corridor, as illustrated in Figure 12.8.

12.4.6 Corridor connections

Where the corridor is used to transfer wind forces horizontally to a stiff core or braced stairwell, the connection of the modules to the corridor may be made by a detail of the form of Figure 12.9. The extended plate is screw fixed on site to the underside of the corridor members and is bolted to the reentrant corners between the four modules, so that it also acts as a tie plate. This detail is not used to provide vertical support to the corridor, which is provided by continuous angles that are preattached to the modules.

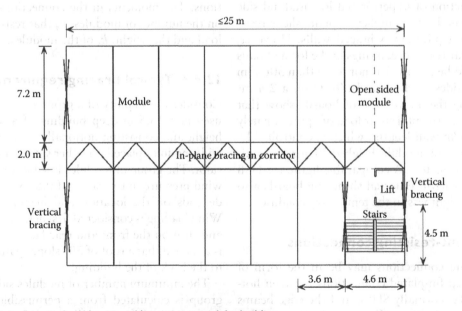

Figure 12.8 Location of vertical and horizontal bracing in a modular building.

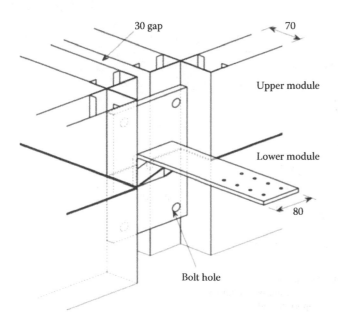

Figure 12.9 Connection between the corridor cassette and modules—sketch detail (left) and actual detail (right). (Courtesy of Unite Modular Systems.)

12.5 EFFECT OF CONSTRUCTION TOLERANCES ON STABILITY

Limits on the geometric deviations of steel-framed structures are presented in BS EN 1090-2: *Execution of Steel Structures and Aluminium Structures*. Technical requirements for steel structures are also given by the British Constructional Steelwork Association (BCSA, 2007). No guidance is given for modular construction for which eccentricities in axial loading have to be considered in the design of load-bearing walls and corner posts in modular construction. The following argument is presented for the potential magnitude of these effects, which is used later in the design of the individual elements.

Lawson and Richards (2010) proposed that the maximum permitted cumulative out-of-verticality at any level due to positional and manufacturing effects may be taken as $\delta_H = 12(n - 1)^{0.5}$ mm for a vertical group of modules, where n is the level considered above the base. This assumes that manufacturing tolerances may act in the same direction as any deviation in installation.

Therefore, the permitted geometric deviation of any pair of modules is 12 mm, when measured from the top of one module to the top of the one below. For the next module, the total out-of-verticality is 17 mm, and so the incremental tolerance is only 5 mm for the next pair of modules. This implies that errors in installation are corrected over the building height. Above the 11th storey, the upper limit of 40 mm on the cumulative out-of-verticality will apply, which means that a much greater control in installation and manufacture is required for taller modular buildings.

12.5.1 Application of notional horizontal forces

The stability of a group of modules due to their potential out-of-verticality is to use the notional horizontal force approach for steel-framed structures given in Clause 2.4.2.3 of BS 5950-1. For steel frames, a horizontal force is applied at each floor level that corresponds to 0.5% of the factored vertical load acting per floor, or 1% of the factored dead load, and is used as a lower-bound alternative to the applied wind force. The 1% limit controls where the self-weight of the floor exceeds its imposed loading, which is the case for modular systems.

BS EN 1993-1-1: Eurocode 3, clause 5.3.2 permits an out-of-verticality of a single column of height/200, but in BS EN 1090-2, this is reduced by a factor of two-thirds when considering the average out-of-verticality over a number of storeys (i.e., an average of $\delta_H \le$ height/300). The permitted out-of-verticality of a whole structure is obtained by multiplying this value for a single column by a factor of

$$\left[0.5 \left(1 + \frac{1}{m} \right) \right]^{0.5}$$

for m columns in a group horizontally. The result for a group of columns tends to be $\delta_H \le$ height/420. A further requirement in BS EN 1993-1-1: Eurocode 3 is that this out-of-verticality is considered in combination with wind loading rather than as an alternative load case, which is the approach in BS 5950-1.

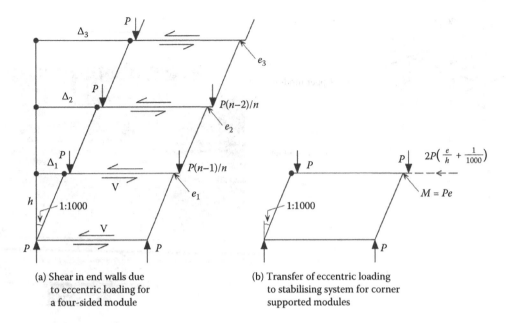

(a) Shear in end walls due to eccentric loading for a four-sided module

(b) Transfer of eccentric loading to stabilising system for corner supported modules

Figure 12.10 Combined eccentricities acting on the ground floor modules.

The combined eccentricity of a vertical assembly of modules takes into account the effects of eccentricities of one module placed on another, but also the reducing compression forces on the walls acting at the increased eccentricity with height. This effect is illustrated in Figure 12.10. The walls of the module are unable to resist high moments, and so the equivalent horizontal forces required for equilibrium are transferred as shear forces in the ceiling, floors, and end walls of the modules.

The additional moment acting on the base module due to the combined effect of the eccentricities of loading in manufacture and installation may be represented by an effective eccentricity Δ_{eff}, given by

$$M_{\text{add}} = P_{\text{wall}}\Delta_{\text{eff}} = P_{\text{wall}} \tag{12.1}$$

$$\left[\frac{(n-1)}{n} \times 12 + \frac{(n-2)}{n} \times 17 + \ldots + \frac{12(n-1)^{0.5}}{n}\right]$$

where P_{wall} is the compression force at the base of the ground floor module, Δ_{eff} is the effective eccentricity of the vertical group of modules, and n is the number of modules in a vertical assembly.

As a good approximation, it is found that the following formula holds for the effective eccentricity of the vertical group of modules, Δ_{eff}, which is used to determine the overturning moment on the base of the building:

$$\Delta_{\text{eff}} = 3n^{1.5} \text{ mm for } n < 12 \text{ storeys} \tag{12.2}$$

This eccentricity may be converted to a notional horizontal force applied at each floor level, which is expressed as a percentage of the factored load at each floor level. Using the tolerances defined above, it is calculated that the notional horizontal force varies with the number of storeys according to $0.2n^{0.5}\%$, when expressed as a percentage of the vertical load applied to a module. For $n = 12$ storeys, it follows that the notional horizontal force per floor is 0.7% of the factored loading on a module.

For modular construction, it is recommended that the notional horizontal force is taken as a minimum of 1% of the factored vertical load acting on each module, and this is used as the minimum horizontal load in assessing overall stability of the structure.

For a module of 25 m² floor area supporting a factored loading of 8 kN/m², the notional horizontal force acting in either direction of the module is 2 kN. For 10 modules in a vertical group, the base shear is therefore 20 kN per module. This force may be shared between the two walls of the module in the direction of the force. The combined effect of the notional horizontal forces may exceed the wind force on the end gable when there are more than seven modules in a horizontal group. If the modules are unable to resist this horizontal force required for overall stability, then the notional horizontal forces must be combined for a number of modules on plan at each level and transferred to the stabilising system. This is the case for open-sided modules, which are generally unable to resist this shear force through bending of the corner posts and their connections.

12.6 DESIGN OF STRUCTURAL ELEMENTS

There are six basic components of a load-bearing module whose structural performance depends on whether vertical loads are transferred through the side walls or corner posts. These components are

- Load-bearing side walls
- Non-load-bearing end walls (or side walls if edge beams are used)
- Floors
- Ceiling
- Edge beams
- Corner posts, usually angles or square hollow sections

The design of the cold-formed steel components is carried out to BS 5950-5 or BS EN 1993-1-3. The design of the hot-rolled steel components, usually the corner posts and, in some cases, the edge beams, is carried out to BS 5950-1 or BS EN 1993-1-1.

The design of modular units should include other issues of structural performance, such as

- Diaphragm action for transfer of in-plane forces due to wind actions and notional horizontal forces
- Structural integrity or robustness to accidental actions, which are generally resisted by tying forces developed at the connections
- Fire resistance, which requires that the structural members are fire protected and that fire does not spread from one module to another

The structural design of the components of a module is presented as follows.

12.6.1 Load-bearing walls

The load-bearing walls in four-sided modules consist of lipped C sections that are 70 to 100 mm deep and are manufactured in steel thicknesses of 1.5 to 2.0 mm in steel strengths S350 to S450 (350 to 450 N/mm² yield strength). The vertical C sections are placed at 400 or 600 mm centres along the wall, or in pairs at 600 mm centres for heavily loaded walls. Plain C section "tracks" form the top and bottom of the prefabricated wall panels and transfer the vertical load between the walls of the modules above and below.

The C sections are designed to resist compression based on their effective length, ℓ_{eff}, between the floor and ceiling. Buckling in the plane of the wall is essentially prevented by fixing of the C sections to the internal plasterboard and external sheathing boards.

Therefore, the effective slenderness of the C sections where boards are attached on both sides of the wall is given by

$$\lambda = \ell_{eff}/r_{yy} \qquad (12.3)$$

where r_{yy} is the major axis radius of gyration of the C section (using axes defined in the Eurocodes).

For walls with no external sheathing boards, but with two layers of plasterboards attached on one side of the C section, the effective length, ℓ_{eff}, for minor axis buckling may be taken as half of the wall height provided that the walls are X-braced for stability. This leads to a lower compression resistance than the case where sheathing boards are fixed on the outside of the wall.

The compression strength of the C section depends on its initial imperfection over its height, and is obtained from the Perry-Robertson strut buckling formula. This is explained by reference to BS EN 1993-1-1, which is expressed in terms of a non-dimensional slenderness ratio, $\bar{\lambda}$. The compression strength of the C section is given by $p_c = \chi f_y$, where f_y is the steel design strength. The value of the reduction factor due to buckling, χ, is calculated for the slenderness ratio, $\bar{\lambda}$, according to

$$\chi = \frac{1}{\Phi + \sqrt{\Phi^2 - \bar{\lambda}^2}} \quad \text{but } \chi \leq 1,0 \qquad (12.4)$$

where

$$\Phi = 0,5\left[1 + \alpha\left(\bar{\lambda} - 0,2\right) + \bar{\lambda}^2\right] \qquad (12.5)$$

and the slenderness ratio is given by

$$\bar{\lambda} = \lambda / \lambda_1 \text{ where } \lambda_1 = \pi\sqrt{E/f_y} \qquad (12.6)$$

E is the elastic modulus of steel, which is 210 kN/mm². α is an imperfection factor given in Table 12.4, which corresponds to the appropriate buckling in BS EN 1993-1-1. For cold-formed sections, buckling curve b should be used, in which case $\alpha = 0.34$.

The compression resistance of the member, P_C, is calculated from

$$P_C = A_{eff} \cdot \chi f_y \qquad (12.7)$$

where A_{eff} is the effective area of the cross section, which takes into account local buckling of the flat plates of the C section, and A_{eff} is typically in the range of 0.7 to 0.9 times the unreduced cross-sectional area, A_{gross}.

An additional moment is considered in the design of the C sections, which is taken as the axial load in the wall multiplied by an eccentricity given by the maximum tolerance in the placement of the module above. In Section 12.4, the nominal eccentricity of 12 mm is

Table 12.4 Imperfection factors for buckling curves for steel columns to BS EN 1993-1-1

Buckling curve	a_o	a	b	c	d
Imperfection factor α	0.13	0.21	0.34	0.49	0.76

considered in the placement of the modules, and this is recommended as the minimum eccentricity to determine the moment acting on the C sections in combination with axial compression. Lateral-torsional buckling of the C section does not occur in walls, when restrained by plasterboards and sheathing boards. Therefore, the bending resistance may be taken as the elastic bending resistance of the C section in its major axis direction. Minor axis bending is not considered.

Bending and compression acting on the C sections in the side walls are combined according to

$$\frac{P}{P_C} + \frac{Pe + M_w}{M_{el}} \leq 1.0 \qquad (12.8)$$

where P is the compression force acting on the C sections in the wall, e is the eccentricity in load application (taken as a minimum of 12 mm), M_w is the bending moment in the C section due to wind acting on the face of modules on the gable of the building, P_C is the compression resistance of the C section, based on buckling in the major axis direction, and M_{el} is the elastic bending resistance of the C section in the major axis.

Wind loading and vertical loading are combined using the load factors given in Section 12.1. It is generally found that this equation is satisfied when the maximum compression load, P_{max}, is approximately 60% of the buckling resistance, P_C, of the C section.

Properties of 100 mm deep by 50 mm wide C sections that may be used to determine the load-carrying capacity of light steel walls in modular construction are presented in Table 12.5. For example, the compression resistance of a 100 mm deep by 1.6 mm thick C section is 36 kN for a wall height of 2.7 m. Therefore, pairs of C sections placed at 600 mm centres can resist up to 120 kN/m length of the wall. This compression resistance is equivalent to the wall loads at the base of a 12-storey building.

Table 12.5 Typical bending and compression resistances of C sections with boards attached on both sides of the wall

C section (S350 steel)	Bending and compression resistances		
	M_{el}kNm	P_CkN	P_{max}kN
70 × 50 × 1.6 mm C	2.3	36	25
100 × 50 × 1.4 mm C	3.2	38	27
100 × 50 × 1.6 mm C	3.8	51	36
100 × 50 × 1.8 mm C	4.6	69	48
100 × 50 × 2 mm C	5.4	89	62

Note: Data for effective wall height of 2.7 m and for C sections in S350 steel. P_{max} is the maximum compression resistance allowing for 12 mm eccentricity of axial load.

12.6.2 Floors

Floors of modules support the imposed and dead loads applied directly to them. The prefabricated floors generally comprise 150 to 200 mm deep C sections in 1.2 to 1.5 mm thick steel at 400 mm spacing. The C sections in floors are placed at 400 mm centres to be compatible with the floorboards that are used. Lateral-torsional buckling is prevented by the attachment of the floorboards to the top flange. The design of these C sections in floors is generally controlled by deflections or perceptible vibrations rather than their bending resistance.

The effective bending properties of the C section are also influenced by local buckling, although to a lesser extent than for members in pure compression. It is necessary to check that under the load combinations in Table 12.1,

$$M \leq M_{el} \qquad (12.9)$$

where M is the bending moment acting on the floor joist.

The serviceability limits that apply to the design of light steel floors are taken from SCI P301 (Grubb et al., 2001), as follows:

Imposed load deflection	≤ Span/450
Total load deflection	≤ Span/350 but ≤15 mm
Natural frequency, f	≥ 8 Hz for rooms
	≥ 10 Hz for corridors and communal space

These limits are stricter than for steel beams in order to ensure that the lightweight floors feel "stiff." The natural frequency limit is used in order that the effects of rapid walking do not lead to perceptible vibrations. This is ensured by designing for a natural frequency of the floor of at least three times the maximum walking pace for lightweight floors. A simple check on the natural frequency is to use the formula

$$f = 18/\sqrt{\delta_{sw}} \text{ Hz} \qquad (12.10)$$

where δ_{sw} is the deflection (in mm) due to the self-weight of the floor and an additional load of 30 kg/m², which is considered to be the permanent component of the imposed load in residential buildings. It follows that $\delta_{sw} \leq 5$ mm (for the 8 Hz limit) and $\delta_{sw} \leq 3.2$ mm (for the 10 Hz limit). Where the dead load is approximately one-third of the total working load, the 15 mm limit on the total load deflection is also satisfied when designing to these frequency limits.

In some cases, a joisted floor of a module is also designed to support a concrete floor screed of 70 to

120 mm thickness, which may incorporate embedded water pipes for heating. In this case, the joist thickness may be increased to 2 mm in order to support the additional floor load of 2 to 3 kN/m².

12.6.3 Ceiling

The ceiling members of the module are designed to support the self-weight of the ceiling itself, and loads applied to it during installation. It is proposed that this temporary construction load is taken as a minimum of a 1 kN/m². This means that the ceiling joists should generally be a minimum of 100 mm depth for spans up to 3.3 m. Often they are chosen as the same size as the floor joists (i.e., 150 or 200 mm), so that the same production system is used for both. This is an advantage where the ceiling of the top module supports the roof and is designed to support snow loading and the self-weight of the roofing.

12.6.4 Edge beams

Edge beams span between corner posts or, in some cases, between the corner posts and an intermediate post. Edge beams are normally provided at the floor and ceiling level, and these beams support the loading transferred from half of the floor or ceiling width (i.e., 1.6 to 2.1 m).

The design principles for edge beams follow that outlined for floors, although the loading and span are higher. The natural frequency criterion generally controls, and so mid-span deflection of the edge beam under the self-weight of the supported floor should be less than 5 mm. For acceptable serviceability performance of edge beams, the ratio of span:section depth should be in the range of 18 to 24. Typically, a 300 mm deep parallel flange channel (PFC) can span up to 7.2 m, depending on the module width. Cold-formed sections and PFCs of up to 430 mm depth may be used for longer spans.

Furthermore, the connections between the edge beams and corner posts are often designed to resist bending moments in order to provide some lateral stiffness for partially or fully open-sided modules. This is generally achieved by fin-plated connections welded to the corner post and bolted to the web of the edge beam. However, this type of connection rarely provides more than 20% of the bending capacity of the edge beam itself.

12.6.5 Corner posts

Corner posts add to the compressive resistance of a wall, and for partially or fully open-sided modules, all the applied vertical loads are resisted by these posts. The posts are partially restrained from buckling by the in-plane stiffness of the modules, but this assumption may not be valid for highly perforated walls. Conservatively, the effective length of the corner posts is taken as the clear distance between the floor and ceiling of the module.

Corner posts are usually in the form of square hollow sections (SHSs) of 100 or 150 mm width. The design of the corner posts is affected by

* Lateral restraint provided by the adjacent walls
* Eccentricity in load application from the module above
* Connections between the modules above and below

In the design of the corner posts, it is recommended that the moment considered to act in combination with the compression load transferred from the modules above is calculated from a minimum eccentricity of 12 mm due to permitted installation and manufacturing tolerances plus the moment due to the load transferred from the attached edge beam at each floor, as illustrated in Figure 12.11. This moment potentially acts in both directions so that biaxial moments have to be considered in the design of the posts.

Assuming that the total compression load on the post is n times the load transferred from one edge beam, the total eccentricity of the axial load acting on the corner post is given by $e = 12 + b/n$ (in mm), where b is the width of the post and n is the number of storeys. For

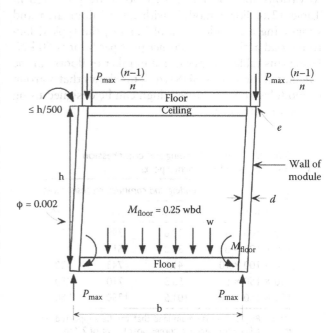

Figure 12.11 Illustration of eccentricity of forces applied to the walls or corner posts of a module.

$b = 100$ mm and $n = 6$, it follows that e is approximately 30 mm. It is recommended that this is taken as the minimum eccentricity used in the design of the corner posts, irrespective of the building height.

Compression and bending actions on the corner posts are combined, as follows:

$$\frac{P}{P_c} + \frac{Pe + M_w}{M_{by}} + \frac{Pe}{M_{bz}} \leq 1.0 \qquad (12.11)$$

where the terms are as described for a module wall and P_c is the axial compression resistance of the post $= \chi\, P_y$, M_{by} is the buckling resistance moment in the y direction, M_{bz} is the buckling resistance moment in the z direction, e is the effective eccentricity of compression load, taken as a minimum of 30 mm in both the x and y directions, and M_w is the bending moment due to wind acting on the corner posts.

Wind loading is considered separately in the x and y directions and is combined with axial load using the load factors in Table 12.1.

For a SHS post that is unrestrained in its height except at its ends, its compression resistance is given by $P_c = A\chi\, f_y$, where χ is the buckling reduction factor calculated from the slenderness of the post in its weaker direction. The only difference with respect to the method presented for C sections in walls is that the imperfection parameter for SHS members is given as for buckling curve a in Table 12.4.

Typical compression loads, P_{max}, that may be applied to various sizes of SHS corner posts are presented in Table 12.6. For a module with 25 m² floor area and supporting a corridor area of 2.5 m², the typical factored load acting on the corner post per floor is 60 kN. Using this table, the permitted number of floors can be calculated for a given SHS size. This shows that a group of modules up to 16 storeys high can be designed using 150 × 150 SHS corner posts.

Table 12.6 Typical bending and compression resistances of SHS corner posts

SHS section (S355 steel)	Bending and compression resistances		
	M_{by}kNm	P_CkN	P_{max}kN
100 × 100 × 5	23.6	398	270
100 × 100 × 8	34.8	613	430
100 × 100 × 10	41.1	743	520
150 × 150 × 5	55.3	710	490
150 × 150 × 10	101.5	1380	950

Note: P_{max} is the maximum load that can be supported by the post. Data for effective corner post height of 2.7 m.

12.7 STRUCTURAL INTEGRITY

12.7.1 General requirements

Robustness, which is also called structural integrity, is concerned with stability and localisation of damage in accidental or extreme loading events, as required by the Building Regulations Approved Document A (2004). One way of satisfying this requirement is to provide for alternative load paths by adequate tying action between the elements of the construction. Guidance on tying requirements in BS 5950-5 refers to the principles in BS 5950-1 in which the tying force is not less than the vertical shear force acting on the element. Guidance on the robustness of light steel framing and modular construction was given by Lawson et al. (2005) and is explained as follows.

In modular construction, the way of assessing the stability of the group of modules is to consider notional removal of one support to the corner of a module and to ensure that the effect of damage to the module is localised. For these calculations, a reduced imposed load factor of 0.33 and a dead load factor of 1.05 may be considered. Wind loading may be ignored for this extreme design case.

Modular units are generally tied horizontally and vertically at all four corners, as illustrated in Figure 12.12. These connections are made through plates and single bolts, which are installed sequentially as each module is placed. Tying at an internal junction of a group of modules can prove problematic, as illustrated in this figure. The fourth module to be placed cannot be easily connected at its base unless access to the connection is through the service riser or another opening.

12.7.2 Robustness of modular systems

For modular construction, robust structural action may be established by considering various scenarios for localisation of damage, corresponding to loss of support at the ground or intermediate floor. Figure 12.13 shows two extreme cases of loss of a corner support or an intermediate support due to notional removal of part of a ground floor module. This corresponds to loss of a corner support or, alternatively for continuously supported modules, loss of support to one end and half of a long side of the module.

The forces due to loss of this support are resisted by tying forces between the modules. It may be assumed that the ties to each module resist the loads applied to that module. The modules themselves are inherently robust in terms of their manufacture, and the forces

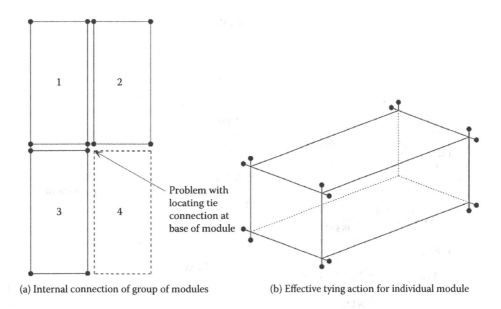

(a) Internal connection of group of modules   (b) Effective tying action for individual module

Figure 12.12 Tying action between modules.

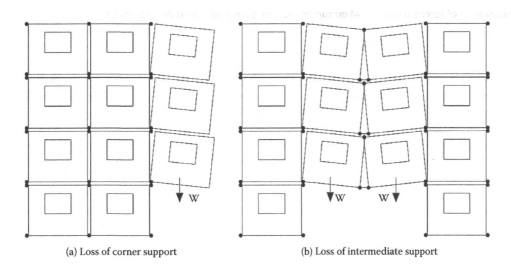

(a) Loss of corner support   (b) Loss of intermediate support

Figure 12.13 Robustness scenarios in modular construction.

developed due to removal of one support can be resisted by in-plane forces in the walls, which are braced or sheathed by various types of boards.

Figure 12.14 shows the results of a finite element analysis of a module when one support is removed and which takes account of the torsional and bending stiffness of the module due to diaphragm action of the walls (Lawson et al., 2008). In this analysis, the maximum horizontal tying force that was developed was 26% of the total load applied to the module. Therefore, it is recommended that the minimum value of the horizontal tying force may be taken safely as one-third of the total load applied to a module in this condition (i.e., for 1.05 × self-weight of the module plus one-third of the design-imposed load).

For lightweight modules with a self-weight of up to 4 kN/m² and a floor area of 25 m², it follows that the minimum tying force is approximately 35 kN. For heavier modules with a self-weight up to 6 kN/m², the minimum tying force should be increased to 50 kN.

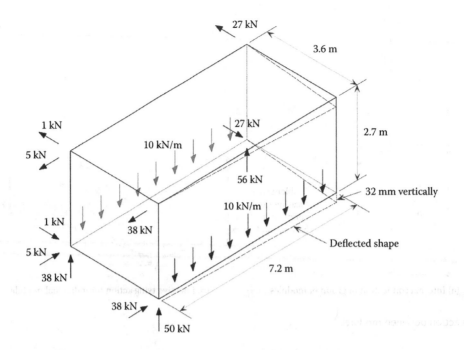

Figure 12.14 Illustration of forces in the ties when support to one corner of a module is removed.

CASE STUDY 40: ELEVEN-STOREY STUDENT RESIDENCE, TOTTENHAM

Transport of two study bedroom modules per lorry.

X-bracing in the corridors for stability.

Unite Modular Solutions' fourth high-rise building was constructed as part of Hale Village, which is an urban regeneration project next to the main line Tottenham Hale station in north London. The 680-bedroom student residence is 11 storeys high, cascading to 5 storeys. In the taller part of the diamond-shaped building, nine floors of modules sit on a podium structure consisting of two open-plan communal levels.

The innovative part of this design was the installation of braced light steel corridor cassettes to transfer wind loads from the vertical stack of modules to the cores at the four corners of the building, which are supplemented by X-braced shear walls. The proximity to the main railway line meant that the precaution was taken to mount the modules on neoprene strip bearings.

The modules varied in external width from 2.7 m for the standard study bedrooms to 3.7 m for the microflats. Kitchen modules were 3.4 m wide. The walls of the modules consist of 70 × 1.6 mm C sections placed at 300 to 600 mm spacing, depending on the load applied at a given level. The floors use 150 × 1.6 mm C sections.

Two tower cranes were able to install modules at a rate of one floor of 60 modules every 2 weeks from podium level using an eight-man module installation team, which meant that the upper floors were completed three to four times faster than in situ construction. The project was completed in only 11 months from ground slab, saving an estimated 12 months.

The modules are tied together at their reentrant corners using square plates with four M16 bolts fixed to welded nuts. The same connector plates also provide an attachment for steel angles, which are screw fixed to the X-braced light steel corridor cassettes (see above).

Modules were installed to very high accuracy by using lasers passing from the base slab passing through holes in the external face of the modules. In this way, positional accuracy of less than 3 mm could be achieved. The ground floor modules were installed on 50 mm thick neoprene-bearing strips with a 4 mm steel-bearing plate above. The horizontal position of the modules was adjusted to take account of tolerances in the adjacent concrete core. These deviations were accommodated in the 20 mm gap between the modules.

The building gained a BREEAM "Very Good" rating, and a U-value of less than 0.22 W/m2K was achieved in the façades by the insulated render and other forms of lightweight cladding.

REFERENCES

British Constructional Steelwork Association. (2007). *National structural steelwork specification for building construction.* 5th ed. London.

British Standards Institution. (1996). *Loading for buildings. Code of practice for dead and imposed loads.* BS 6399-1.

British Standards Institution. (1997a). *Structural use of steelwork in building. Part 5. Code of practice for design of cold formed thin gauge sections.* BS 5950-5.

British Standards Institution. (1997b). *Loading for buildings. Part 2. Wind loading.* BS 6399-2.

British Standards Institution. (2000). *Structural use of steelwork in building. Part 1. Code of practice for design in rolled and welded sections.* BS 5950-1.

British Standards Institution. (2004a). *Steel structures—General rules and rules for buildings.* BS EN 1993-1-1: Eurocode 3.

British Standards Institution. (2004b). *Specification for continuously hot-dip zinc coated structural steel and strip—Technical delivery conditions.* BS EN 10346.

British Standards Institution. (2004c). *Gypsum plasterboards. Definitions, requirements and test methods.* BS EN 520.

British Standards Institution. (2005a). *Actions on structures. Part 1-1. General actions—Densities, self-weight, imposed loads for buildings.* BS EN 1991-1-1: Eurocode 1.

British Standards Institution. (2005b). *Actions on structures—General actions. Part 1-4. Wind actions* (and its UK national annex). BS EN 1991-1-4: Eurocode 1.

British Standards Institution. (2006). *Steel structures—General rules. Supplementary rules for cold-formed members and sheeting.* BS EN 1993-1-3: Eurocode 3.

British Standards Institution. (2008). *Execution of steel structures and aluminium structures. Technical requirements for steel structures.* BS EN 1090-2 and amendment A1 (2011).

Building Regulations (England and Wales). (2004). Structure: Approved document A. www.planningportal.gov.uk.

Grubb, P.J., Gorgolewski, M.T., and Lawson, R.M. (2001). *Building design using cold formed steel sections. Light steel framing in residential construction.* Steel Construction Institute P301.

Lawson, R.M. (2007). *Building design using modules.* Steel Construction Institute P367.

Lawson, R.M., Byfield, M., Popo-Ola, S., and Grubb, J. (2008). Robustness of light steel frames and modular construction. *Proceedings of the Institute of Civil Engineers: Buildings and Structures,* 161(SB1).

Lawson, R.M., Ogden, R.G., Pedreschi, R., Popo-Ola, S., and Grubb, J. (2005). Developments in pre-fabricated systems in light steel and modular construction. *Structural Engineer,* 83(6), 28–35.

Lawson, R.M., and Richards, J. (2010). Modular design for high-rise buildings. *Proceedings of the Institute of Civil Engineers: Buildings and Structures,* 163(SB3), 151–164.

Chapter 13

Structural design of concrete modules

This chapter covers the structural design of concrete modules and addresses some of the important aspects related to the layout of the modules based on their structural performance. Relevant requirements for the design of reinforced concrete to BS EN 1992: Eurocode 2 are also presented. Issues related to the construction and installation of concrete modules are presented in Chapter 17, and they also affect the structural design of the modules.

13.1 DESIGN PRINCIPLES IN MODULAR PRECAST CONCRETE

In modular precast concrete construction, the reinforced concrete walls transfer the vertical and horizontal loads through the structure to the foundations. The use of concrete can also help to meet additional requirements of

- Fire resistance
- Acoustic separation
- Concealed services distribution (electrical and sanitary)
- Internal and external finishes to walls
- Thermal mass to assist in controlling internal temperatures

In addition, modular construction using precast concrete also has the following structural features:

- A shallow floor zone (150 to 200 mm depending on the floor span), where the ceiling of one module forms the floor of the one above.
- Thin walls (125 to 150 mm), although a double wall is formed by adjacent modules.
- A pair of rooms can be accommodated within one module.
- The load capacity of the walls is very high, making tall buildings feasible.
- Horizontal stability is provided by the walls of the modules.

The structural design of precast concrete modules is the same as the design of in situ reinforced concrete.

The module dimensions should be standardised wherever possible, allowing the precast manufacturer to fully utilise available moulds in the factory production. The modules should be designed and configured to make the casting, striking, lifting, and installation as simple as possible.

13.2 CONCRETE PROPERTIES

The controlling factor in concrete mix design is usually the concrete strength at de-moulding. The units have to be "struck" from the formwork, generally 12 to 24 h after casting, and moved so that the next unit can be cast. Due to this rapid turnaround in the manufacturing process, various methods are used to accelerate the early-age strength of the concrete, such as use of rapid hardening cement, chemical accelerators, and external heating, such as steam curing and electrical heating, either outside or inside the formwork. Typical strengths of concrete required at de-moulding and at 28 days are given in Table 13.1. The typical class of concrete used in precast modules is C35/45 (cylinder/cube strength in N/mm^2); see later for definitions of concrete properties.

Precast concrete manufacturers modify their mix designs depending upon the local supplies and, in particular, the grading of the aggregates. To make efficient and economic use of the casting moulds, manufacturers usually use higher strength concrete than would generally be used for in situ concrete.

Self-compacting concrete (SCC) is also increasingly used in precast concrete to reduce the use of vibrators, and hence labour to compact the concrete. SCC also generally has a higher strength and superior surface finish when compared to conventional concretes (Goodier, 2003).

13.3 CODES AND STANDARDS

BS EN 1992: Eurocode 2: *Design of Concrete Structures* is the relevant design code for reinforced concrete members, including precast concrete. Eurocode 2 co-existed with BS 8110 for many years, but as of 2010, national

Table 13.1 Typical strengths of precast concrete

Component	Nominal grade	Cube strength at 28 days (N/mm²)	De-mould cube strength (N/mm²)	Design strength (N/mm²)	Elastic modulus at 28 days (kN/mm²)
Beams, shear walls, floors	C30/40	40	20–25	18.0	28
Columns, load-bearing walls	C40/50	50	25–30	22.5	30

Source: Modified from Elliott, K. S., *Precast Concrete Structures*, BH Publications, Poole, UK, 2002.

Note: Concrete strengths are specified in cylinder/cube strength in N/mm².

codes have been withdrawn. The two parts of Eurocode 2 that are used in the design of concrete building structures are

- BS EN 1992-1-1: *Common Rules for Buildings and Civil Engineering Structures*
- BS EN 1992-1-2: *Structural Fire Design*

Each part has a national annex (NA) that gives national values for certain partial factors (or nationally determined parameters (NDPs)). Advice on the new codes is available from www.eurocode2.info, www.eurocodes.co.uk, and the Concrete Centre. A design manual has been published by the Institution of Structural Engineers (2006).

13.3.1 EN 13369 and other product standards

Precast concrete elements should also conform to the appropriate product standard, EN 13369: *Common Rules for Precast Concrete Products*, and the relevant standards are listed in Table 13.2. They are written "by exception" to EN 13369; i.e., they either accept what is in EN 13369 or have mirror clauses that supersede those in the EN standard.

The main chapters refer to the areas of application and to materials, manufacture, manufacturing tolerances, minimum dimensions, concrete cover, surface

Table 13.2 Relevant BS EN standards for precast concrete

Standard	Title
BS EN 1168:2005	Hollow-Core Slabs
BS EN 12794:2005	Foundation Piles
BS EN 13224:2004	Ribbed Floor Elements
BS EN 13225:2004	Linear Structural Elements
BS EN 13369:2004	Common Rules for Precast Concrete Products
BS EN 13693:2004	Special Roof Elements
BS EN 13747:2005	Floor Plates for Floor Elements
BS EN 14650:2005	General Rules for Factory Production Control of Metallic Fibred Concrete
BS EN 14843:2007	Stairs
BS EN 14991:2007	Foundation Elements
BS EN 14992:2007	Wall Elements

quality, and resistance to mechanical actions, i.e., the load-bearing capacity. Other parts deal with fire resistance, acoustic insulation, durability, safety in transport and erection, and safety in use.

13.3.2 Eurocode 2: Design of concrete structures

The design of reinforced concrete to Eurocode 2 is based on the characteristic cylinder strength rather than cube strength of concrete, and is specified according to BS 8500: *Concrete*, which is the complementary British Standard to BS EN 206-17 (BSI, 2000). As an example, for class C35/45 concrete, its cylinder strength is 35 N/mm² and its cube strength is 45 N/mm². Typical structural properties of concrete are shown in Table 13.3.

All Eurocodes use limit state design principles in which partial factors are applied to loads (actions) as well as to the material strengths. Partial factors for loads are common to all materials and are presented in Table 12.1. For loads acting in combination, the load factors applied to the variable actions are reduced. For design at the serviceability limit state, unfactored loads are used.

Material properties are specified in terms of their characteristic values, which in general correspond to a defined fractile of an assumed statistical distribution of the property considered (usually the lower 5% fractile). For design at the ultimate limit state in Eurocode 2, the partial factors for materials are $\gamma_c = 1.5$ for concrete and $\gamma_s = 1.15$ for reinforcement. The design strength of concrete in compression is given by $f_{cd} = \alpha_{cc}f_{ck}/\gamma_c$, where f_{ck} is the cylinder strength of concrete, and α_{cc} is a factor that takes account of long-term effects on the compressive strength (α_{cc} is taken as 0.85 in the UK). γ_c is taken as 1.0 at the serviceability limit state.

The simplified stress blocks used in Eurocode 2 are shown in Figure 13.1, which are used to develop the design equations for bending of reinforced concrete walls and floors, and are similar to those found in BS 8110. The maximum concrete compressive stress is taken as $0.85f_{ck}/1.5$, and the depth of the compression block is taken as 80% of the depth of the neutral axis for concrete strengths, $f_{ck} \le 50\text{N/mm}^2$.

Table 13.3 Concrete properties to Eurocode 2-1-1 (BS EN 1992-1-1)

Symbol	Description	Properties (in units or as defined)								
f_{ck} (N/mm²)	Characteristic cylinder strength	12	16	20	25	30	35	40	45	50
f_{ck} cube (N/mm²)	Characteristic cube strength	15	20	25	30	37	45	50	55	40
f_{ctm} (N/mm²)	Mean tensile strength	1.6	1.9	2.2	2.6	2.9	3.2	3.5	3.8	4.1
E_{cm} (kN/mm²)	Secant modulus of elasticity	27	29	30	31	33	34	35	36	37

Note: Mean secant modulus of elasticity at 28 days for concrete with quartzite aggregates.

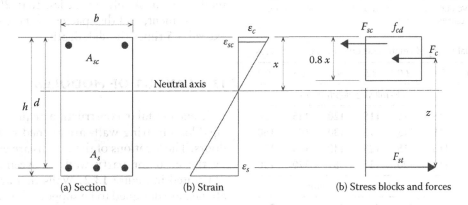

(a) Section   (b) Strain   (b) Stress blocks and forces

Figure 13.1 Simplified stress blocks for reinforced concrete sections in bending.

The design strength of reinforcement is given by $f_{yd} = f_{yr}/\gamma_s$, where f_{yr} is the characteristic strength of the reinforcement (normally 500 N/mm²) and γ_s is the partial factor defined as above. The properties of steel reinforcement in the UK, for design to Eurocode 2, are given in BS 4449: *Specification for the Reinforcement of Concrete* (2005b).

13.3.3 Minimum dimensions of concrete elements

The minimum dimensions of concrete elements are dependent mainly on their fire resistance. This also includes the minimum axis distance (cover plus half bar diameter) to the reinforcement. The minimum width and depth of beams, columns, and slabs are presented in Table 13.4.

Minimum wall thicknesses from a manufacturing point of view are usually in the range of 140 to 170 mm. This is consistent with a fire resistance of 60 or 90 min for walls exposed to fire on both sides. However, for separating walls, a minimum thickness of 180 mm is often used for acoustic reasons. Walls generally contain two layers of steel mesh reinforcement. Care should be taken if the element is designed with large openings or box-outs, especially if near to the end of the wall. The module and its lifting points should also be designed for the forces when it is lifted from its formwork and later during installation (see Chapter 17).

The design of concrete walls in compression is dependent on their effective height. For a wall in a modular unit that is cast monolithically into a ceiling or floor slab at each of its ends, its effective height may be taken as 0.9 times the actual wall height. The moments acting at the ends of the wall should also be considered in combination with the axial load. For scheme design, the wall height should be less than 20 times its width, so that the effects of buckling for a slender wall do not reduce its load-bearing capacity by more than 50% relative to its pure compression resistance.

The maximum span:depth ratio of slabs and beams is dependent on their end fixity, and typical cases are presented in Table 13.5. A 170 mm deep reinforced concrete slab would typically span up to 4.2 m if cast monolithically with the walls. Initial sizing of solid concrete slabs may be obtained using the data in Table 13.6. As the span and the imposed load increase, a thicker slab

Table 13.4 Minimum dimensions of concrete elements (mm) for fire resistance

Element	Fire resistance (min)			
	R30	R60	R90	R120
Columns—fully exposed	200	250	300	450
Columns—partially exposed	155	155	155	175
Walls—exposed on two sides	120	140	170	220
Walls—exposed on one side	120	130	140	160
Beams—width	80	150	200	200
Slabs—depth	150	180	200	200
Axis distance of reinforcement—slabs	20[a]	20[a]	30	40

[a] Practical minimum.

Table 13.5 Maximum span:depth ratios for acceptable serviceability performance of slabs and beams

Element	Span:effective depth
Simply supported beam	14
Continuous beam	18
Simply supported slab	20
Continuous flat slab	24
Cantilever	6–8

Note: Effective depth is from the top of the member to the centre of the tensile reinforcement.

Table 13.6 Initial sizing of reinforcement in solid floor slabs

Single span (m)	2.5	3.0	3.5	4.0	4.5	5.0
Imposed load	Overall slab depth (mm)					
1.5 kN/m²	115	115	115	120	135	150
2.5 kN/m²	115	115	115	130	145	160
5.0 kN/m²	115	115	125	140	160	175
7.5 kN/m²	115	120	135	155	170	190
Imposed load	Reinforcement (kg/m²)					
1.5 kN/m²	3	3	4	4	6	7
2.5 kN/m²	3	4	5	6	7	8
5.0 kN/m²	3	5	6	7	8	11
7.5 kN/m²	4	5	6	7	8	10

Source: Brooker, O., and Hennessy, R., Residential Cellular Concrete Buildings: A Guide for the Design and Specification of Concrete Buildings Using Tunnel Form, Cross-Wall, or Twin-Wall Systems, CCIP-032, Concrete Centre, London, UK, 2008.

with more reinforcement is required. These values are based on the use of C35/45 concrete grade and S500 reinforcement with 20 mm cover, and apply for 60 min of fire resistance (Brooker and Hennesey, 2008).

13.3.4 Reinforcement

Walls and slabs will generally contain two layers of reinforcement, usually in the form of mesh reinforcement, as it is quicker and simpler to fix than individual bars. Early-age thermal and shrinkage effects should also be considered when crack control is important, especially if an exposed finish is required.

The minimum area of vertical reinforcement in the wall should be at least 0.2% of the gross cross-sectional area of the wall, which is equally divided between the two faces of the wall. For effective placement of the concrete, the maximum area of vertical reinforcement should not exceed 4% of the gross cross-sectional area of the wall. Horizontal reinforcement should be provided parallel to the faces of the wall and should have a minimum area equal to either 25% of the vertical reinforcement or 0.1% of the gross cross-sectional area, whichever is greater. The spacing between these bars should not exceed three times the wall thickness or 400 mm, whichever is lower.

For floor slabs, the minimum area of main reinforcement is calculated as a function of the bending moment acting on it due to imposed loading and its self-weight (see Table 13.6). The limit on reinforcement area is the same as for walls. The spacing of main reinforcement should generally not exceed three times the depth of the slab or 400 mm, whichever is greater, and the spacing should be reduced to two times the depth in areas of concentrated loads. The area of secondary (distribution) reinforcement should not be less than 20% of the main reinforcement, and the spacing of these bars should not exceed 3.5 times the slab depth.

13.4 LAYOUT OF MODULES

Efficient modular construction requires that the principal load-bearing walls are aligned vertically between floors. The locations of these walls are usually governed by the need for party walls between apartments, as illustrated in Figure 13.2. Walls that are not vertically aligned are designed to be supported by floors or ceiling of the module. These non-load-bearing walls can still be produced using the same precast modular construction techniques, or can be lightweight infill walls.

Repetition of the manufacture of the modules and their details can achieve significant time savings and material cost. A range of layouts can be achieved through careful use of three or four basic module designs in varied combinations. However, more complex and less repetitive building layouts may result in increased formwork and labour costs, material use and wastage, and longer installation time.

Precast concrete modules can be combined with other systems as required, such as precast slab or wall panels or with steel frames, for example, if a traditional pitched and tiled roof structure is required. Building heights using precast concrete modules can range from single storeys up to 5 or 6 storeys.

13.5 DETAILED DESIGN

After the form and layout of modules has been agreed on, the detailed design is carried out. Some factors to be considered in detailed design are presented as follows.

13.5.1 Robustness and stability

Modular precast concrete structures are very resistant to lateral loads due to the large number of load-bearing walls. However, in the design of open-ended modules, stability may be more problematic where lateral loads act perpendicular to the walls. Also, temporary stability during construction should be considered for open-ended modules.

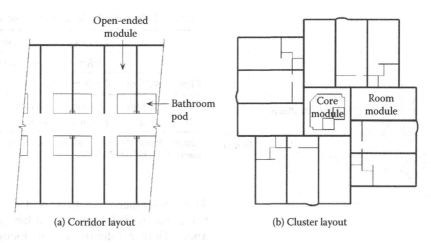

(a) Corridor layout   (b) Cluster layout

Figure 13.2 Typical layouts for modular precast structures: (a) linear cross-wall arrangement and (b) arrangement around a central core. (From Brooker, O., and Hennessy, R., *Residential Cellular Concrete Buildings: A Guide for the Design and Specification of Concrete Buildings Using Tunnel Form, Cross-Wall, or Twin-Wall Systems*, CCIP-032, Concrete Centre, London, UK, 2008.)

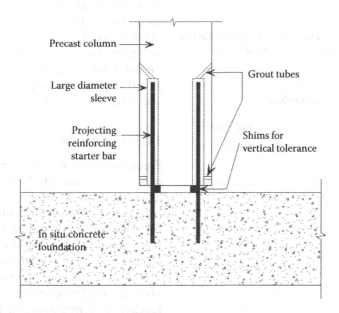

Figure 13.3 Projecting starter-bars for foundation connections. (From Concrete Centre, *Precast Concrete in Buildings*, Report TCC/03/31, London, UK, 2007.)

Continuous concrete walls and slabs are inherently robust and can easily meet the requirements for structural integrity by appropriate reinforcement detailing. In cross-wall construction, the joints between the individual panels and slabs should be sufficiently reinforced for tying action.

Method statements demonstrating temporary stability of the modules during construction should also be prepared.

13.5.2 Joints and connections

A number of different methods of connecting precast concrete modules to other elements may be used. These connections should be able to transmit forces in three directions between the structural elements. Joints within the modules and connections to adjacent modules must also be capable of providing structural integrity.

To connect concrete modules and other precast elements to in situ concrete foundations, projecting starter-bars are often cast into the foundation, as shown in Figure 13.3. The precast units can then be craned onto the foundation, and the starter-bars are inserted into holes in the units. The module is aligned and placed onto steel shims in order to achieve the correct line and level. The joints are then grouted when the modules are in their correct position and are levelled. In situ concrete may also be placed on the precast units to form the final surface of the floor, and the minimum depth of this topping is normally 60 mm.

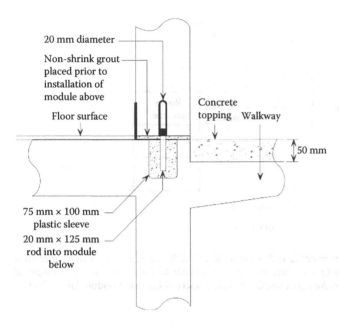

Figure 13.4 Internal connection of a precast concrete prison unit with a balcony. (Courtesy of Oldcastle Precast.)

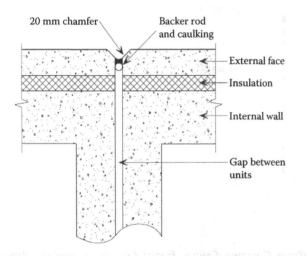

Figure 13.5 Exterior connection in precast concrete prison (plan view). (Courtesy of Oldcastle Precast.)

Figure 13.4 shows a similar principle for placing and connecting one precast concrete module on top of another. Steel bars and coils are grouted into box-outs in the module in order to fix them together. A plan view of two precast concrete modules is shown in Figure 13.5, in which insulation is incorporated within the external faces of the module. The two modules are not structurally connected to one another, and the gap is sealed with backer rod and caulking.

13.5.3 Tolerances

Recommended production tolerances are given in the product standards for precast concrete (see earlier).

Table 13.7 Permitted tolerances for wall elements

| | Permitted deviation in wall element | | | | |
| | Reference dimensions of element | | | | |
Class	<0.5 m	0.5–3 m	3–6 m	6–10 m	>10 m
A	±3 mm	±3 mm	±3 mm	±3 mm	±10 mm
B	±8 mm	±14 mm	±16 mm	±18 mm	±20 mm

Source: British Standards Institution, *Precast Concrete Products: Wall Elements*, BS EN 14992, 2007.

These tolerances can be varied in the final specification, and the values presented here are given for guidance. Tighter tolerances may incur a cost premium in manufacture.

The permitted deviations of lengths, heights, thicknesses, and diagonal dimensions of wall elements are shown in Table 13.7, which is taken from BS EN 14992. This standard has two classes for tolerances: class A, being generally more onerous than BS 8110, and a less stringent class B. Class A is more likely to be used for modular buildings due to the implications of dimensional inaccuracy on the overall structure. Floor tolerances are provided in BS EN 13747.

13.5.4 Balconies

Precast concrete balconies are manufactured mainly for use in residential buildings or hotels, and can be cast as part of the module or attached at a later date. The balcony units are cast with reinforcement projecting from the back that can be connected to the reinforcement in the concrete module. This may take the form of threaded bars placed into pre-prepared sockets or holes, which are later grouted on site. The balcony units are supported temporarily until the grout or screed has reached sufficient strength. The balcony units typically incorporate integral drainage details and an up-stand to facilitate proper weatherproofing at door interfaces. Tiled or finished upper faces may also be incorporated, together with cast-in fittings for up-stands or handrails.

Balconies or walkways can also be cast monolithically as part of the ceiling of the main module, as shown in Figure 13.6.

13.5.5 Foundations and transfer structures

Wide strip footings or piled ground beams can be used to support concrete modules. Strip footings are only appropriate for low-rise precast concrete structures. For piled foundations, the reinforced concrete ground beams transfer the loads to the piles. The layout and weight of the modules and the ground conditions influ-

Figure 13.6 Balcony cast into a precast module. (Courtesy of Oldcastle Precast.)

ence the final pile design. The column loads can be many hundreds of tonnes, depending on the number and size of the modules supported.

Concrete modules can also be supported by transfer structures above open-plan areas. Because of the loads that are supported, the transfer structure usually takes the form of a relatively thick reinforced slab supported by a series of columns, whose spacing is chosen depending on the use of the space below. These columns transfer the imposed loads and self-weight of the modules onto the foundations, usually in the form of pile groups and pile caps located under the walls or columns.

13.5.6 Design requirements for installation

Installation is a key aspect of the use of precast concrete units. Design considerations for installation include the crane capacity, both at the precast yard and on site, transport, and access to the site. Precast concrete construction requires sufficient space on site for the delivery, unloading, and storage of the modules. Mobile cranes or sometimes overhead gantry cranes are used in the factory. Figure 13.7 shows a module being installed and the temporary support to the base of the open module. This is considered further in Chapter 17.

Figure 13.7 Installation of precast concrete modules on site showing the temporary support to the base of the walls. (Courtesy of Oldcastle Precast.)

REFERENCES

British Standards Institution. (2000). *Concrete. Specification, performance, production and conformity.* BS EN 206-1.

British Standards Institution. (2004a). *Common rules for precast concrete products.* BS EN 13369.

British Standards Institution. (2004b). *Design of concrete structures. Part 1-1. Common rules for buildings and civil engineering structures.* BS EN 1992-1-1, Eurocode 2.

British Standards Institution. (2004c). *Design of concrete structures. Part 1-2. General rules—Structural fire design.* BS EN 1992-1-2, Eurocode 2.

British Standards Institution. (2005a). *Precast concrete products. Floor plates for floor systems.* BS EN 13747.

British Standards Institution. (2005b). *Steel for the reinforcement of concrete—Weldable reinforcing steel—Bar, coil and decoiled product—Specification.* BS 4449.

British Standards Institution. (2006). *Concrete. Parts 1 and 2.* BS 8500 (complementary British Standard to BS EN 206-1).

British Standards Institution. (2007). *Precast concrete products: Wall elements.* BS EN 14992.

Brooker, O., and Hennessy, R. (2008). *Residential cellular concrete buildings: A guide for the design and specification of concrete buildings using tunnel form, cross-wall or twin-wall systems.* CCIP-032. Concrete Centre, London, UK.

Concrete Centre. (2007). *Precast concrete in buildings.* Report TCC/03/31. London, UK.

Concrete Centre. (2008a). *Properties of concrete for use in Eurocode 2.* London, UK.

Concrete Centre. (2008b). *How to design concrete buildings to satisfy disproportionate collapse requirements.* London, UK.

Concrete Centre. (2009a). *Concrete and the code for sustainable homes.* London, UK.

Concrete Centre. (2009b). *Design of hybrid concrete buildings.* London, UK.

Concrete Centre. (2011). *How to design concrete structures to Eurocode 2—The Compendium.* London, UK.

Elliott, K.S. (2002). *Precast concrete structures.* BH Publications, Poole, UK.

Goodier, C.I. (2003). The development of self-compacting concrete in the UK and Europe. *Proceedings of the Institution of Civil Engineers: Structures and Buildings,* 156(SB4), 405–414.

Institution of Structural Engineers. (2006). *Manual for the design of concrete building structures to Eurocode 2.* London, UK.

Narayanan, R.S. (2007). *Precast Eurocode 2: Design manual.* CCIP-014. British Precast Concrete Federation, Leicester, UK.

Chapter 14

Cladding, roofing, and balconies in modular construction

Cladding systems may be pre-attached to the module, or installed as separate elements on site. In both cases, the connection between the modules may be concealed or emphasised as part of the detailing of the cladding. The features of the various types of cladding and their influence on the design of the modules are presented in this chapter. In precast concrete modules, the external face of the module may be finished in the factory, as described in Chapter 15.

The thermal performance of cladding systems and the integration of renewable energy technologies are both aspects of interest in modern design, which are also covered in this chapter.

14.1 CLADDING TYPES FOR LIGHT STEEL MODULES

Four generic forms of cladding may be considered in the design of modular buildings using light steel framing:

1. Ground-supported brickwork, in which the brickwork is constructed conventionally on site from the foundation or podium level and is laterally supported by the modules.
2. Insulated render that is applied on site to insulation that is fixed to the external sheathing boards of the modules. This type of lightweight cladding is supported by the modules and conceals the joints between the modules.
3. Rain screen cladding systems in the form of boards, tiles, or metallic sheets that are fixed through insulation to the external sheathing boards. For heavier tiled systems, horizontal rails are attached through to the light steel structure of the modules.
4. Brick slips attached to metallic sheets or bonded to sheathing boards that are attached through insulation to the modules. Although the brick joints are mortar filled on site, this type of cladding system is not generally considered to be weathertight. Also, its weight adds to the loads acting on the modules.

Brickwork is generally designed to support its own self-weight up to 3 or 4 storeys' height (approximately 12 m). Lateral support is provided by the modular units by brick ties connected to stainless steel or corrosion-protected vertical runners that are screw fixed at 600 mm centres through the external sheathing boards to the light steel framework of the modules. The brick ties are attached normally every fifth brick course, or every third course around windows. Details of brickwork attachments are shown in Figure 14.1.

It is not normal practice for the light steel modules to provide vertical support to brickwork unless an additional steel support structure is provided. Lintels are required over window openings in the brickwork. However, brickwork cladding is often used for the lower level of a building and lightweight cladding is used above.

Lightweight cladding takes many forms, from insulated render and tiles to metallic sheets and cementitious boards. These cladding systems may be designed to be supported entirely by the modules over any building height. The use of insulated render on a separate sheathing board is illustrated in Figure 14.2. The sheathing board provides weather resistance in the temporary condition and improves the airtightness of the building in service.

In the case of rain screen cladding systems, horizontal or vertical rails are screw fixed to the light steel framework of the modules through the external insulation and sheathing boards. When the thickness of the closed-cell insulation board exceeds about 100 mm, the fixings may become too flexible and do not support the tiles or boards effectively. In this case, separate stainless steel or aluminium L-shaped brackets may be required, which can be adjusted to allow for the site tolerances in the placement of the modules.

For all rain screen cladding systems, the modules are designed to be weathertight and to provide the required level of thermal performance, independent of the type of cladding that is used. For metallic cladding systems, either vertical or horizontal rails may be pre-fixed to the light steel framework of the modules, as illustrated in Figure 14.3. This system acts as a rain-screen, and

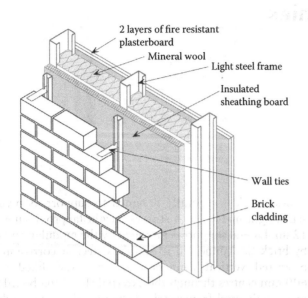

Figure 14.1 Typical brickwork cladding attachment to a light steel substructure.

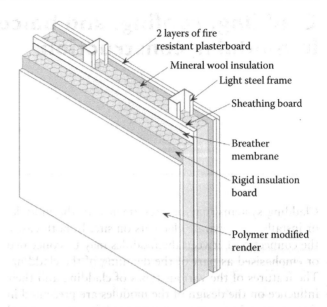

Figure 14.2 Typical rendered cladding attached to sheathing board.

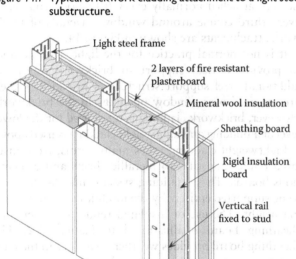

Figure 14.3 Steel cassette cladding with perforated C sections in this case.

Figure 14.4 Tiles supported by composite panels. (Courtesy of Kingspan.)

so an additional sheathing board is required. In this figure, the C sections are shown as perforated, which reduces the effect of thermal bridging through the structural elements.

Another form of metallic system is to use horizontally spanning composite panels (also known as sandwich panels), which provide greater rigidity and dispense with the need for an insulation board. Composite panels are weathertight and are designed to add to the thermal insulation of the façade. It is possible to connect tiles to composite panels via horizontal rails that are fixed to the outer steel sheet of the panel, as shown in Figure 14.4.

Examples of these types of cladding used on modular buildings are illustrated in Figures 14.5 to 14.10.

14.2 THERMAL PERFORMANCE

The use of modular construction is increasing in housing and residential buildings, as its off-site nature of construction leads to more reliable thermal performance of the building envelope. The important thermal performance characteristics are

- Thermal insulation
- Minimising thermal bridging
- Airtightness
- Control of condensation

Thermal insulation is characterised by the thermal transmission or U-value of the building envelope, which is the measure of the rate of heat loss through 1 m² of the

Figure 14.5 Ground-supported brickwork up to 4 storeys high and insulated render above for a modular residential development in Basingstoke.

Figure 14.6 Innovative insulated render and metallic cladding to a student residence in north London. (Courtesy of Unite Modular Solutions.)

Figure 14.7 Bonded brick tile cladding to a student residence in north London. (Courtesy of Unite Modular Solutions.)

Figure 14.8 Rain screen cladding to social housing in east London. (Courtesy of Rollalong.)

Figure 14.9 Metallic cladding to conference centre in Leamington Spa. (Courtesy of Terrapin.)

envelope for 1° difference in temperature across it, and its units are W/m²K. Representative U-values to achieve the energy use targets in current regulations were given earlier in Table 6.2 (Zero Carbon Hub, 2009).

U-values less than 0.2 W/m²K for external walls and 0.15 W/m²K for roofs are generally specified in current projects. This requires use of technologies with proven thermal performance in which factors such as thermal bridging and airtightness are properly considered.

Loss of heat by air infiltration through the building envelope can be responsible for over 30% of the total heating requirement in a modern well-insulated building. Therefore, it is equally important to improve airtightness as to improve thermal insulation. This often requires the use of airtight membranes or well-jointed sheathing systems. However, as the airtightness of buildings increases, it also is necessary to maintain fresh air quality and to eliminate the risk of condensation by use of controlled ventilation systems with heat recovery.

Light steel construction uses the "warm frame" framing principle, where the majority of insulation is placed externally to the structure, as shown in Figures 14.1 to 14.3. In modular construction, additional mineral wool insulation is placed between the C sections in the walls for acoustic insulation and fire resistance purposes. It is recommended that to avoid any risk of interstitial condensation, at least two-thirds of the total insulation level that is provided should be placed externally to the frame. This is satisfied when at least 60 mm of closed-cell insulation board is placed outside the light steel frame, because it has a lower thermal conductivity than

the 100 mm thick mineral wool placed between the C sections.

14.2.1 Thermal properties of common building materials

The thermal transmittance through a unit surface area of a material is defined by its thermal conductivity, λ, divided by its thickness, d. Metals have relatively high thermal conductivities, whereas insulation materials, such as mineral wool and polyurethane, are good insulators and have low thermal conductivity. Thermal conductivities of common building materials are presented in Table 14.1. For a multilayer wall or roof, the thermal resistances d_i/λ_i of the various layers (subscript i) are added together to determine their combined resistances. The U-value is the inverse of the total thermal resistance of the element. Surface resistances may be added to take account of local heat transfer at internal and external surfaces and of internal cavities, but these effects are relatively small in terms of their effect on the U-value.

14.2.2 Control of thermal bridges

A thermal bridge occurs when any component made of a material with high thermal conductivity leads to

Table 14.1 Thermal properties of building materials

Material	Thermal conductivity λ-value ($Wm^{-1}K^{-1}$)	Thermal resistance (m^2KW^{-1})
Steel	50	
Stainless steel	16	
Aluminium	160	
Plasterboard	0.25	
Render	1.0	
Wood or timber boarding	0.17	
Brickwork	0.77	
Blockwork—heavy	1.44	
Blockwork—lightweight	0.19	
Concrete	1.65	
Extruded/expanded polystyrene (EPS)	0.032–0.035	
Mineral wool	0.037–0.040	
Polyurethane (PUR)/ polyisocyanurate (PIR) closed-cell insulation	0.025	
Air gap (high emissivity)		0.18
Air gap (low emissivity)		0.44
External surface resistance		0.04
Internal surface resistance		0.13

Figure 14.10 Mixed use of insulated render, rain screen cladding, and ground-supported balconies in modular apartments in Dublin.

higher heat flow in comparison to the adjacent surface area of lower thermal conductivity. The majority of thermal bridges occur at points of discontinuity in the structure, such as in corners, joints, windows, and doors, and at gaps in the insulation. In severe cases, moisture condensation caused by thermal bridging can affect the long-term durability of the building.

Repeating thermal bridges should be included explicitly in the calculation of the U-value of the façade system. Additional linear thermal bridges should be calculated separately in the form of Ψ-values, and are added into the overall heat loss calculation. These Ψ-values are multiplied by the length of the thermal bridge, and divided by the exposed area of the wall in order to establish their overall effect in terms of heat loss. Point thermal bridges may also occur where beams or balconies penetrate the building envelope. The total influence of these thermal bridges might be as high as 20% of total transmission loss.

14.3 THERMAL PERFORMANCE OF LIGHT STEEL MODULAR WALLS

The thermal performance of a light steel wall with 100 mm deep C sections at 600 mm spacing is presented in Table 14.2 for an insulated render system with various types of insulation—expanded polystyrene (EPS), polyurethane (PUR), or polyisocyanurate (PIR). In addition, mineral wool is placed between the

Table 14.2 U values for insulated render to a light steel wall without a cavity

Insulation Thickness (mm)	Expanded Polystyrene EPS ($\lambda = 0.035$ W/mK)	Closed-Cell Insulation PIR/PUR ($\lambda = 0.025$ W/mK)
60	0.27	0.23
80	0.23	0.19
100	0.20	0.16
120	0.18	0.14

Source: Lawson, R.M., Sustainability of Steel in Housing and Residential Buildings, The Steel Construction Institute, P370. Note: Mineral wool is placed between the C sections in all cases. U-value takes account of the C sections in the wall.

C sections, a sheathing board is attached externally, and a single layer of plasterboard is attached internally.

The thermal profile through the wall is shown in Figure 14.11, which illustrates the local heat loss through the C sections. A U-value of 0.2 W/m²K is achieved for an insulated render system with 100 mm of expanded polystyrene (EPS) or 80 mm of closed-cell (PIR/PUR) board bonded externally to the sheathing board.

14.4 AIRTIGHTNESS

Airtightness of a building envelope significantly influences the energy consumption of the building. The air leakage points of a building are concentrated mainly at the junction of building components and service connections.

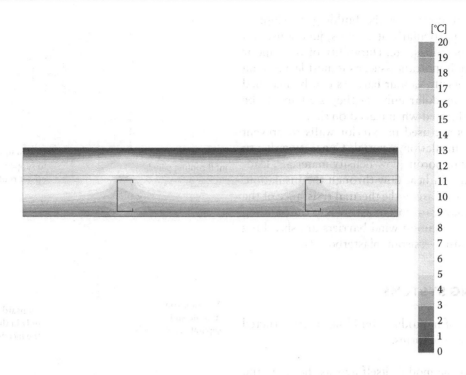

Figure 14.11 Thermal profile in insulated render system to light steel wall (80 mm thickness of polyurethane insulation externally). (Data by the Steel Construction Institute.)

14.4.1 Definition and measurement methods

Airtightness is generally expressed in terms of an air leakage value, which is defined by air pressure testing of the building. It is measured using a blower door test, which consists of a calibrated fan for measuring an airflow rate and a pressure sensing device to measure the pressure created by the fan. The combination of pressure and flow can be used to estimate the airtightness of the building envelope.

Airtightness should be tested according to BS EN 13829 at 50 Pa pressure difference between inside and outside. Air leakage may be defined as air changes per hour (ach), which is the volume of air entering the building relative to its enclosed volume in 1 h, and is expressed in $n50[h^{-1}]$. Alternatively, it may be represented by the volume of air contained within the building divided by the surface area of the building envelope, again expressed as over 1 h (m^3/m^2h). This factor is called the air permeability, or q_{50}. The ratio between these two parameters depends on the scale and proportions of the building.

A typical single-family house has a surface area of 200 to 250 m^2 and a volume of 300 to 350 m^3. The ratio of ach to m^3/m^2 is such that 1 ach is approximately equivalent to 1.5 m^3/m^2h air leakage through the building envelope in a single-family house. This ratio will be different for other building sizes and forms.

For the UK, the target air permeability value for buildings is 10 $m^3/m^3/h$ (measured at a pressure of

Table 14.3 Airtightness data based on measurements for typical buildings

Building type	Air leakage rate at 50 Pa ($m^3/m^2/h$)
Typical on-site construction as envisaged by the Building Regulations	10
Residential building using modular construction	1–3
Prefabricated timber or light steel-framed houses (terraced house)	3–5
Low-energy detached house	0.8–2

50 Pa). All new buildings over 500 m^2 floor area must be airtightness tested and the actual value used in energy calculations. For buildings that are not tested, a default value of 15 $m^3/m^3/h$ must be used in building energy calculations. Data-measured results of the airtightness of typical buildings are presented in Table 14.3. A modern modular building can achieve high levels of airtightness, which is significantly better than the equivalent on-site construction.

14.4.2 Influence of vapour and wind barriers

Vapour barriers play a very important role in controlling condensation. The primary requirement for a vapour barrier is its water vapour resistance as it prevents warm, humid air from intruding into the colder building envelope. Continuity in the vapour barrier over the internal surface of the building envelope is very important, particularly at corners, junctions, and details around openings, etc. Durability of the vapour barrier is also an important issue, as it must last as long as the building itself. Vapour barriers can be installed more reliably in modular units, as they are liable to be damaged or perforated when placed on site.

Wind barriers are used in exterior walls to prevent airflow into the insulation material. Convection due to wind will occur in porous, low-density materials, leading to an increase in heat flow through the insulation. It will lead to a decrease in the thermal resistance of the insulation and may transport moisture into the building fabric. The most common wind barriers are sheathing boards and moisture-resistant plasterboards.

14.5 ROOFING SYSTEMS

Generally, the roof in modular buildings is constructed in one of four generic forms:

1. The top of the module itself acts as the roof, and it is weatherproofed and laid to falls, often with integral downpipes in the corners of the modules.
2. Purlins that span parallel to the building façade and support a pitched roof. The purlins may be attached to triangular or curved wall frames that are positioned over the load-bearing side walls of the modules.
3. Roof trusses that span perpendicular to the building façade and are supported by the front and rear façade walls of the modules (or the corner posts of the modules).
4. Modular roof units that are designed to create habitable space. In this case, the modules are generally of mansard shape and are supported directly by the modules below.

Examples of the different types of roof system used in modular construction are illustrated in Figure 14.12. In this case, a set-back module is supported on the side walls of the modules below, but the roof balcony is supported by the ceiling of the module below.

A curved roof was used in a recent modular housing project in west London, as shown in Figure 14.13. This was achieved by a separate curved light steel structure that was supported by the modules.

A mansard roof may be constructed using modules that are manufactured with the required roof profile. A good example is shown for a five-student residence in York (see Figure 14.14), in which the set-back roof was required for planning reasons.

In all cases, the interface between the roof and the modular units is designed to resist both compression due to gravity loads and tension due to wind uplift on

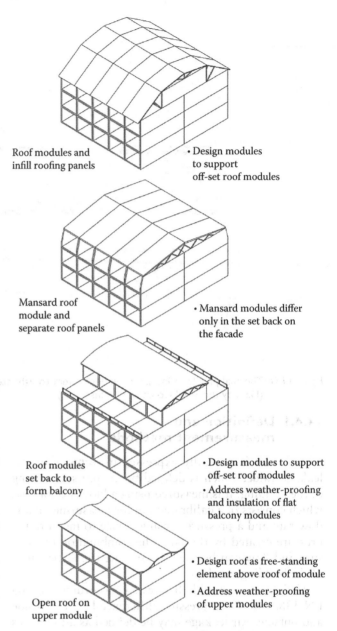

Roof modules and infill roofing panels
- Design modules to support off-set roof modules

Mansard roof module and separate roof panels
- Mansard modules differ only in the set back on the facade

Roof modules set back to form balcony
- Design modules to support off-set roof modules
- Address weather-proofing and insulation of flat balcony modules

Open roof on upper module
- Design roof as free-standing element above roof of module
- Address weather-proofing of upper modules

Figure 14.12 Examples of roof structures in modular construction.

the roof. The magnitude of these forces will increase depending on the span and pitch of the roof. In the case of a 12 m span roof with 6° slope, uplift forces may be of the order of 10 kN per holding-down point.

14.6 BUILDING IN RENEWABLE ENERGY TECHNOLOGIES IN MODULAR CONSTRUCTION

Renewable energy technologies may be integrated into modular units or may be attached to the roof and walls of modular buildings. The most common renewable energy solutions are photovoltaic panels and solar thermal collectors.

Figure 14.13 Curved house roofs in the Birchway project, west London. (Courtesy of Futureform.)

Figure 14.15 Photovoltaic panels used in combination with a "green" curved roof for modular housing. (Courtesy of Futureform.)

Figure 14.14 Mansard roof modules used in a 5-storey student residence in York. (Courtesy of Elements Europe.)

14.6.1 Photovoltaics

Photovoltaic cells use semiconductor-based technology to convert light energy into an electric current. The electrical energy that is created can be fed into the national grid (exported) by an inverter that converts the DC to the AC at the mains voltage. There are two forms of photovoltaics (PVs)—either crystalline or more rigid forms that are used to manufacture self-standing panels, or laminates in the form of amorphous silicone that are bonded to a metallic surface.

Large PV panels are generally supported on horizontal rails that are attached to the roof, as shown in Figure 14.15. They are generally located on the south-facing slope of roofs, although east- and west-facing roofs can be used with some loss of efficiency. Dark grey or black tiles incorporating PV layers are attractive, as they do not detract from the visual appearance of a more conventional house. A good example of a traditional house with integral PV tiles is illustrated in Figure 14.16.

The peak power output of a crystalline PV panel is around 20 W/m² panel area in the UK climate. The average yearly output is likely to be around 100 kWh/m² of the roof, taking account of seasonal variations and roof orientations. For a typical family house with 20 m² of south-facing roof, the energy created can be up to 2000 kWh, which is equivalent to about half of the energy required for space heating of a well-insulated house.

Other forms of PVs may be bonded to glass to be used in roofs or solar shading, as shown in Figure 14.17. This type of PV is more appropriate for commercial or, in some cases, educational buildings.

Figure 14.16 Photovoltaic roof tiles used in a traditional house. (Courtesy of Woking Borough Council.)

Figure 14.17 Photovoltaic glazing used as solar shading.

Modular units can be manufactured with attachment points for PV panels, and electrical inverters and cabling may be installed within the module to reduce installation, commissioning, and testing time on site.

14.6.2 Solar thermal collectors

Different kinds of solar thermal collectors have been used for many years. In modern systems, sunlight is converted into heat through an absorber where water-glycol solution circulates as the heat transferring liquid. The warmed liquid is transferred to a water boiler that utilises the heat energy to assist in heating household water. The system is controlled by a pump and control unit. Solar thermal systems are also used to provide lower-temperature heating in under-floor heating systems.

Solar thermal collectors are usually stand-alone units that are attached to the roof, but they can also be integrated into metallic cladding systems.

14.6.3 Mechanical ventilation systems

Highly insulated and airtight buildings require effective ventilation to avoid buildup of stale air, smells, and high humidity levels. Mechanical ventilation and heat

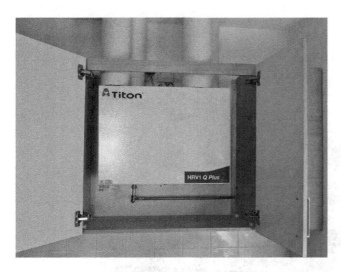

Figure 14.18 Mechanical ventilation and heat exchanger installed in a kitchen unit in a module. (Courtesy of Futureform.)

recovery (MVHR) systems are often introduced by providing extracts in each major room, and particularly in the kitchens and bathrooms. The extracts pipe the warm room air to a heat exchanger (generally located in the loft), which transfers heat to the incoming cooler outside air. Modular units can be manufactured with inbuilt pipes and extracts, and also with MVHR systems located within the modules next to external walls. An example of a small MVHR unit located in a kitchen unit of a module is shown in Figure 14.18.

14.7 BALCONIES

Balconies are an important feature that give greater interest and create useable space in an otherwise bland façade. Prefabricated or integral balconies are therefore an important component of modular construction.

In conventional construction, the floor is continued beyond the building to form the balcony. However, this solution creates a "cold bridge" and does not comply with modern regulations. Also, to minimise the risk of water flowing back into the building, the finished surface of the balcony should be below the internal floor surface.

Balconies can be constructed in various ways in modular systems:

- Ground-supported balconies, which are stacked vertically, and where the columns used to support the modules extend to the ground level.
- An additional external steel structure, which is braced and self-stabilising, and is therefore independent of the modular building.

- Balconies cantilevering from hot-rolled steel posts that are located in the module walls. Fin plates may project from these posts (which are generally square hollow sections) to permit later attachment of the balconies and to minimise cold bridging.
- Integral balconies manufactured as part of the modules. The balconies have side walls or corner posts in this case and are insulated to prevent heat transfer to the modular unit below.
- Balconies supported between the sides of the adjacent modules. In this way, the modules provide both the vertical and the lateral support to the balconies.
- Suspended balconies by ties from each floor to corner posts.

Figure 14.19 shows the use of ground-supported balconies, in which the balcony is partly supported by posts that extend to the ground. The balconies are tied to the modules to resist wind loads. This is a practical solution in modular construction. Alternatively, a separate steel structure may be introduced to provide overall stability and also to support the balconies, as was done at the MOHO project in Manchester shown in Figure 14.20.

Cantilevered balconies, as illustrated in Figure 14.21, require substantial steel-to-steel connections to resist the applied moments transferred from the balconies. This generally requires use of hot-rolled steel members (normally square hollow sections (SHSs)) that are inbuilt into the modules. The balcony attachments may be made to the SHS posts that are generally located at the corners of the modules. To minimise cold bridging, thermal separators can be introduced in the balcony connections, and the wall insulation is locally applied after the structural connections have been made.

A simpler technique is to manufacture the balcony or external space as part of the module, which has been done in various projects. Here the external space must be made watertight and is generally partially enclosed, as shown in Figure 14.22. The sides of the modules project to form the sides of the balcony. An alternative approach is to suspend balconies between the sides of adjacent projecting modules, as shown in Figure 14.23. Fin plates project from the side of the modules to make the on-site attachments of the balcony.

Tied balconies can be relatively unobtrusive, but they must be tied back to the corner posts of the module, as they apply horizontal forces to their supports. Details of the support of a balcony at a corner post of a module are shown in Figure 14.24.

Figure 14.19 Ground-supported balconies in a modular housing, Malmo, Sweden. (Courtesy of Open House AB.)

Figure 14.20 Balconies supported by a separate structure at MOHO Manchester. (Courtesy of Yorkon.)

Figure 14.21 Balconies supported at corner posts of modules. (Courtesy of Caledonian Modular.)

Figure 14.22 Balconies integrated into the modules at Raines Court, north London. (Courtesy of Yorkon.)

Figure 14.23 Balconies supported between the modules. (Courtesy of Open House AB.)

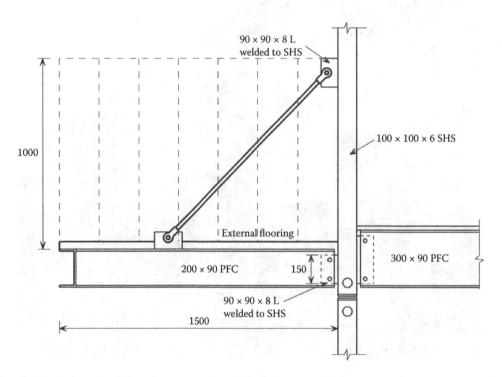

Figure 14.24 Detail of tied balcony to SHS corner posts of modules.

REFERENCES

British Standards Institution. (2001). *Thermal performance of buildings. Determination of air permeability of buildings. Fan pressurization method.* BS EN 13829.

Building Regulations (England and Wales). (2010). *Conservation of fuel and power.* Approved document L1.

Lawson, R.M. (2007). Sustainability of steel in housing and residential buildings. The Steel Construction Institute, P370.

Zero Carbon Hub. (2009). Defining a fabric energy efficiency standard for new homes. www.zerocarbonhub.org.

REFERENCES

Chapter 15

Service interfaces in modular construction

Services located within a module are installed in the factory, and the final service connections to the central services and drainage of the building are made on site. The vertical and horizontal distribution of services throughout the building is an important part of the design process. Service installation is time-consuming in traditional construction and is often on the critical path. In modular construction, a high proportion of the services in the building can be installed and tested off site.

Service interfaces include the connection of the services within the modules to the service distribution within the rest of the building. Services also include lifts and plant rooms, which may also be manufactured in modular form. The general requirements for services in modular buildings are explored in this chapter for both steel-framed and concrete modules.

15.1 SERVICES IN LIGHT STEEL MODULES

In light steel modules, individual wall, floor, and ceiling panels can be manufactured with their own electrical cables, which may be clipped together at the junctions of the panels when the module is assembled. Where possible, services should be designed to run parallel to the primary framing members. However, openings in floor joists and wall C sections may be required, and to prevent fraying of cables, these openings should have rubber grommets around their perimeter when used for electrical distribution.

Vertical service ducts are usually incorporated in the corners of modules, as shown in Figure 15.1. The possible locations for the service risers in a module are illustrated in Figure 15.2, and their features are described as follows. In case (a), the corner post is not connected to the adjacent walls, which means it has to be designed to be structurally stable. In case (b), the service opening is smaller and is formed as part of the wall. In case (c), the service riser is located outside the line of the modules, which means the stability of the corner of the module is not affected by the service opening, but conversely, the

corridor width has to be increased to accommodate the service riser.

Vertical service routing external to the modular units requires fire stopping around ducts and pipes at the floor and ceiling levels to prevent passage of smoke in the event of a fire. The points where the services enter the modules also have to be sealed.

In some types of buildings, it is possible to provide multiple service risers from the central plant, which can reduce the need for horizontal distribution of services. Services, such as chillers, can also be located within the enclosed roof space. In large buildings, services and plant rooms can also be installed in modular form.

The typical service zone in a light steel module is shown in Figure 15.3, in which access to the vertical pipes in adjacent modules is combined in one service riser. The module has no corner post in this case, which would have to be considered in terms of its load-bearing resistance and the lifting method. The fire stopping around the pipes is illustrated in Figure 15.4. The sheathing board in this case should be noncombustible, as it should prevent passage of smoke or flame between the modules through the vertical service zone.

Bathroom pods can be manufactured with thin walls and floors (as little as 50 mm) and installed on the floor of the modules so that the depth of the acoustic floor aligns with the floor of the pod. Features of bathroom pods are described in Chapter 4. Waterproofing below "wet" areas and extending up the walls of the module is also recommended.

Other service strategies that may be used in modular buildings include the following:

- Use of corridors and other spaces for distributing services along the building
- Use of the floor and ceiling voids within each module for distribution of pipes, cables, and air circulation ducts
- Drainage connections to vertical risers in the corner of the modules
- Wet areas connected back-to-back to combine their vertical service zones

Figure 15.1 Service riser in the corner of a module. (Courtesy of Caledonian Modular.)

A further modular option, shown in Figure 15.5, is to manufacture modules that comprise a pair of bathrooms. In this case, the module width can be chosen to be suitable for container transport. Servicing is common to the pair of bathrooms, which reduces the on-site service attachments. In other hybrid forms of construction, discussed in Chapter 10, it is efficient to manufacture bathrooms and kitchens as load-bearing modules and to construct the rest of the structure in panel form.

15.2 SERVICES IN CONCRETE MODULES

Services may be attached to and installed within the concrete modules before delivery to site, as shown in Figure 15.6. Electrical services may be cast in conduits within the concrete itself, which is more visually acceptable and tamper resistant (important for secure accommodation) than surface-mounted electrical distribution. This requires a more integrated approach to procurement and detailed design of the service layout. Additional services, such as under-floor heating, can also be incorporated, if required.

For most precast concrete modules, the water and waste services are distributed vertically to each module or pair of modules. Vertical risers are usually located in the bathroom area in the corner adjacent to the corridor, and therefore provide for maintenance access. It should be decided early in the design process whether a riser is required for each module, or for a pair of modules, as shown in Figure 15.7. In some cases, services can be accessed from the building core, where the modules are clustered around the core.

15.3 MODULAR PLANT ROOMS

There are specialist manufacturers of modular plant rooms that are used in major multistorey office and other buildings. Modular plant rooms include all the central heating and ventilating equipment that is accessible for easy maintenance. The modules are clad when used in external (generally rooftop) applications. Modular plant rooms are often quite large because of the equipment that is also large, and modules can be designed as partially open-sided so that two or more modules create larger plant rooms.

An example of the structure of a modular plant room and its cladding is illustrated in Figure 15.8. In this case, the structure of the modules used small square hollow sections that are welded together. Composite panels were used externally to provide the required watertight enclosure and thermal insulation.

15.4 MODULAR CORES

In large office buildings or hospitals, a number of different types of modules may be used to accommodate the toilets, plant rooms, stairs, and lifts. Toilet modules can be built to a range of dimensions and specifications, and are used to speed up the servicing and fit-out process. In office buildings, the floor depth of the module should be equal to the depth of the raised access floor in the general open-plan space, and this is typically 150 mm. An example of eight modules that link together to provide communal facilities in a combined lift, stair, and bathroom facility is illustrated in Figure 15.9. This system is

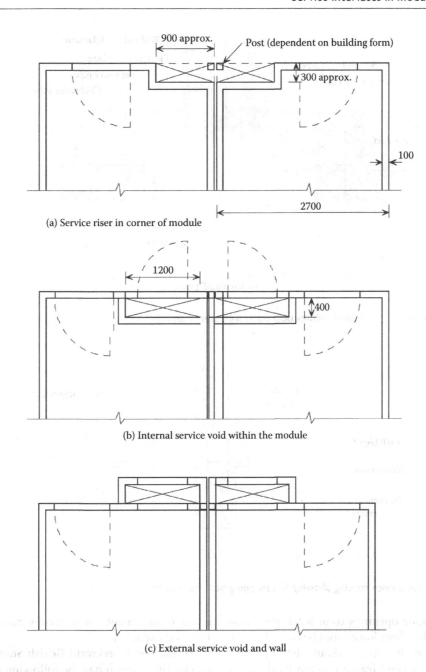

(a) Service riser in corner of module

(b) Internal service void within the module

(c) External service void and wall

Figure 15.2 Possible location of service risers between modules.

more likely to be used in large office buildings in which a braced steel frame is used to provide stability.

15.4.1 Modular lifts

The lift company Schindler first used modular lifts in the late 1980s when there was pressure to install and commission lifts in commercial buildings more rapidly (see Figure 15.10). Recently, modular lifts were installed in Terminal T5 at Heathrow Airport.

Modular lifts permit guide rails, doors, and finishes to be installed in the factory. Guide rails are accurately aligned in the factory to minimise site adjustment. A modular lift is generally constructed using four different module types:

- A lift pit module (1.4 or 1.7 m high)
- A door height module (typical module for a particular lift type)
- A floor zone module (the structural zone is project specific, and this module is used to take account of this zone)
- A capping module (the module at the head of the lift shaft that will generally house the lift motor)

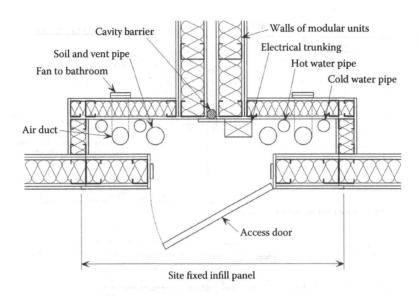

Figure 15.3 Typical service riser between modules built using light steel framing.

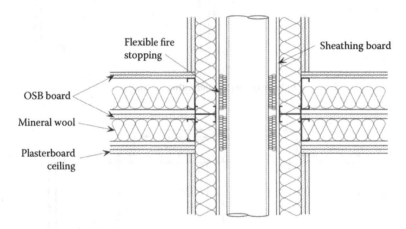

Figure 15.4 Typical vertical service routing showing fire stopping between floors.

Lift shafts have door openings to at least one elevation, and therefore this elevation cannot be cross-braced. Lateral movements of lift shafts should be kept to a minimum, and thus a rigid frame is often used across the door elevation, which has hot-rolled steel posts and crossbeams. Braced light steel infill walls form the three closed sides of the lift shaft. Temporary weatherproofing is provided during construction.

The tolerances required of lift shafts are much tighter than for construction in general, and it is important to eliminate vertical misalignment between the lift shaft modules and the adjacent structure. Shims are used at the corners of the modules to minimise this misalignment.

Since modular lifts were first introduced, the on-site construction of lifts has improved considerably. A lift that would have taken 8 to 10 weeks to install 20 years ago now only takes approximately 2 weeks. Improvements in lift technology have reduced the size of lift-driving equipment to such an extent that there

is no requirement for a motor room for lifts up to 1600 kg capacity.

BS 5655 is the relevant British Standard for lifts and service lifts, which has the following parts:

- Part 5: Has been superseded by BS ISO 4190-1: 1999, which provides information on lift shaft sizes.
- Part 6 (2012) provides guidance on tolerances, types of drive, and standard interfaces.

The position of the guide rails within the shaft varies with the type of lift and details used by a particular manufacturer.

15.4.2 Lift dimensions

In modular construction, the lift shaft may be installed as a part of the main structure. For residential construction, a 630 kg capacity lift is generally adequate for buildings up to 5 storeys high. Figure 15.11 shows

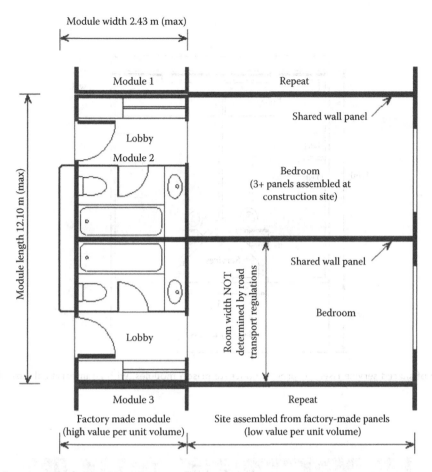

Figure 15.5 Pair of bathrooms manufactured as a single module.

Figure 15.6 Services installed as part of a precast modular unit in manufacture. (Courtesy of Precast Cellular Structures Ltd.)

that such a lift requires clear internal dimensions of 1.9 × 1.6 m. A standard 800 mm door opening is suitable for wheelchair access, increasing to 900 mm for some applications. A deeper lift is required for nursing homes and hospitals where beds are to be moved. The depth of a lift pit in residential construction is 1.4 m.

Lift shafts are often constructed to the standards of separating walls to isolate noise from the lift when in operation. Therefore, it is recommended that two-layer separating walls are provided around lift shafts. The guide rail supports should be provided at a standard dimension of no greater than 2.5 m vertically throughout the shaft, and are subjected to surging and braking loads. This may require use of square hollow section (SHS) ring beams around the lift shaft at floor levels to transfer these effects. The attachment of a modular lift shaft to a light steel supporting structure is shown in Figure 15.12.

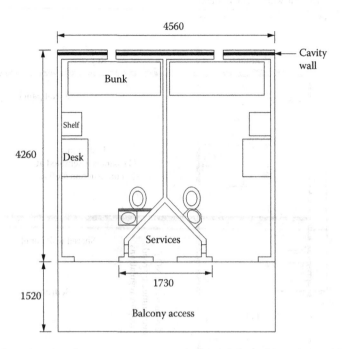

Figure 15.7 Plan view of shared service riser for precast concrete prison modules, with an integral balcony. (Courtesy of Rotondo Weirich.)

(a)

Figure 15.8 Modular plant rooms: (a) bare frame of a module showing its services and (b) clad as a rooftop unit. (Courtesy of Armstrong Integrated Services.)

(b)

Figure 15.8 (continued) Modular plant rooms: (a) bare frame of a module showing its services and (b) clad as a rooftop unit. (Courtesy of Armstrong Integrated Services.)

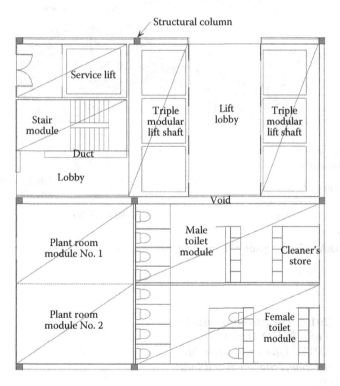

Figure 15.9 Modular lifts, plant rooms, and toilets linked to form a building core.

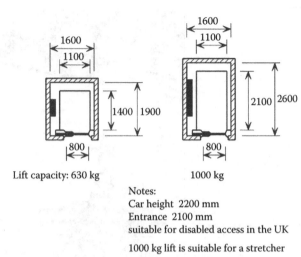

Lift capacity: 630 kg   1000 kg

Notes:
Car height 2200 mm
Entrance 2100 mm
suitable for disabled access in the UK

1000 kg lift is suitable for a stretcher

Figure 15.10 Modular lift system using light steel components. (Courtesy of Schindler.)

Figure 15.11 Lift dimensions to BS ISO 419011.

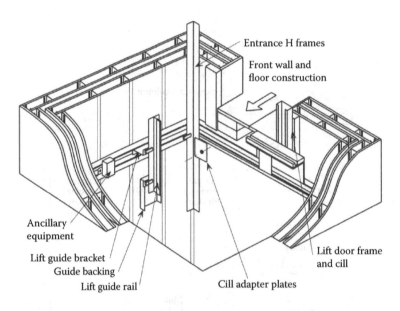

Figure 15.12 Attachment of modular lifts and guide rails in a light steel module.

REFERENCES

British Standards Institution. (2010). *Lift (elevator) installation. Class I, II, III and VI lifts.* BS ISO 4190-1.
British Standards Institution. (2011). *Lifts and service lifts. Code of practice for the selection, installation and location of new lifts.* BS 5655-6.

Chapter 16

Constructional issues in modular systems

The construction of modular buildings requires knowledge of installation methods, connections between the modules, and the interfaces of the modules with foundations, cladding, roofing, and services. Installation rates of 6 to 10 modules per day can be achieved on most sites, depending on weather conditions, site access, travel distances, etc.

This chapter discusses issues of installation processes for both light steel and concrete modules, and their interfaces with the non-modular parts of the building that influence the construction process.

16.1 FOUNDATION INTERFACES

A variety of foundations can be used for modular construction, as shown in Figure 16.1. For modules with load-bearing side walls, strip footings or ground beams supported by pile caps would be the most commonly used foundation systems. For steel modules designed with corner posts, pad footings or pile caps may be used. For concrete modules, where loads are much higher, piled foundations are more often used.

For piled foundations, short spans between each pile lead to smaller ground beams, whereas fewer, larger piles with pile caps lead to longer, and hence deeper, ground beams. A pile cap with three or four piles may be 2 m wide and up to 1m deep, which should be considered when deciding on the ground works. The top surface of the pile cap should be levelled to the required accuracy for placement of the modules by using concrete placed on site.

The accuracy of the foundations and the bearing surfaces on which the modules are placed should be carefully checked before delivery of the modules. The interface between the modular units and the foundations must also provide adequate resistances to forces in all directions in order to satisfy the requirements for structural stability.

For a fitted-out light steel module used in a residential building, the line load acting on the foundation is typically 12 to 20 kN/m (unfactored) per floor level. Therefore, for a 5-storey building, the line load on the foundation may be up to 100 kN/m. For a concrete module, line loads may be 30 to 50 kN/m per floor level.

In corner-supported light steel modules, the concentrated load acting at each corner post is typically 50 to 80 kN (unfactored) per floor. This equates to a total load on the corner post of up to 400 kN in a 5-storey building. In some cases, uplift forces may also act on the foundation system due to wind forces applied to lightweight modules, depending on the plan form and height of the building, and whether an additional bracing system is provided.

The level of module sole plate should be very accurate over the complete module perimeter. The modular units can be levelled using steel shims below the module with a maximum thickness of 20 mm. An acceptable vertical tolerance is 0 to −3 mm over the length of the module. In some systems, vertical pins may be used to locate the modules on the foundations, and also to provide shear resistance. If holding-down fixings to the foundation are required to uplift, they are usually made by chemical anchors.

The interface between the modules and their foundations must provide suitable resistance to moisture to reduce the risk of corrosion. The modules should be located above a design proof course (DPC). Where this is not possible, corrosion protection equivalent to Z460 galvanizing (460 g of zinc per m^2) or a suitable bituminous coating should be applied to all the steel components below the DPC level.

The nature of modular construction is that it provides a suspended ground floor. The Building Regulations Approved Document C (2000) recommends that such floors should have a ventilated air space between the ground and the suspended floor structure, and the ground must be covered with a layer of suitable material to resist moisture. For acceptable heat retention, the ground floor of the modules should also be designed to achieve a U-value of less than 0.15 W/m^{2}K, which may mean attaching additional insulation on top of or below the ground floor module.

A typical detail at the interface between the module and its foundation is shown in Figure 16.2, taken from SCI P302 (Gorgolewski et al., 2001). The sole plate may be in the form of cement particleboard for light loads,

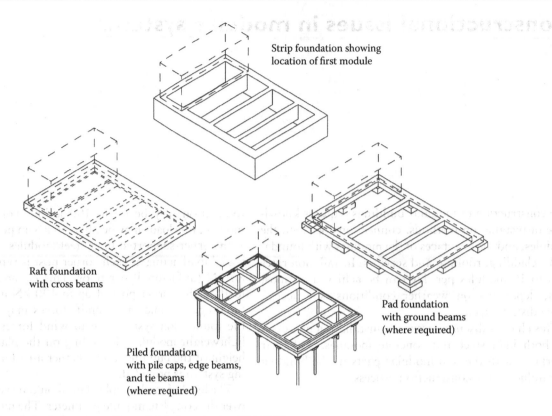

Strip foundation showing
location of first module

Raft foundation
with cross beams

Pad foundation
with ground beams
(where required)

Piled foundation
with pile caps, edge beams,
and tie beams
(where required)

Figure 16.1 Foundation systems used in modular construction (SCI P301).

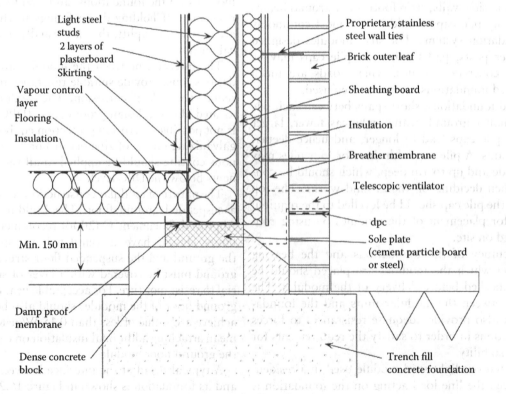

Light steel
studs

2 layers of
plasterboard

Skirting

Vapour control
layer

Flooring

Insulation

Min. 150 mm

Damp proof
membrane

Dense concrete
block

Proprietary stainless
steel wall ties

Brick outer leaf

Sheathing board

Insulation

Breather membrane

Telescopic ventilator

dpc

Sole plate
(cement particle board
or steel)

Trench fill
concrete foundation

Figure 16.2 Typical trench fill foundation supporting brickwork cladding.

or a steel plate for higher vertical loads. It is apparent that the ground floor level of the module may be higher (up to 300 mm) than the external ground level, which should be taken into account when designing for disabled access to the building.

16.2 MODULE TOLERANCES AND INTERFACES

Some degree of dimensional variation may occur due to the manufacturing process for modules of all types. In the case of light steel modules, automated panel fabrication can be very accurate (of the order of +1 to −3 mm), but flexing of components during transport and installation, and the accuracy of placement of the modules, can combine to cause deviations from the theoretical dimensions. In timber-framed construction, further movements can arise from long-term shrinkage, while in concrete construction mould stability and accuracy during casting can have significant effects on the final geometry.

Often the most critical tolerance for low- to medium-rise modular buildings relates to plan dimensions. Any positive dimensional variation reduces the space between modules and can cause clashes between the modules, with lifts and other services, or in extreme cases, incorrect alignment with foundations.

Control of vertical tolerances is more important for buildings of 6 or more storeys' height and is very important for high-rise applications. The differences in the heights of the walls of a module tend to be specific to the system and installation method, but in the extreme case, this can cause stepping out in the height of modules, or vertical out-of-alignment effects where modules become wedge shaped.

Generally, the position of modules can be shifted slightly to maintain the overall verticality of the extremities of the façade, but a sawtooth cross section will result. This may be coupled with problems in maintaining the horizontality of floors in taller buildings, as illustrated in Figure 16.3.

Where vertical load is transferred through the corner posts, shims can be used at the connections between the posts to ensure that the modules are in their ideal positions. Acoustic packs with sufficient stiffness may also be used to provide adjustment, and the thickness of these packs can be varied depending on on-site measured dimensions.

In brick-clad modules, the floor-to-floor height of the modules will normally be designed in units of 75 mm to suit multiple brick courses, and so accurate setting out of the brickwork is required at the foundation level. Other forms of premanufactured cladding require adjustments at the joints that can match the construction tolerances of the modular system. Reconciliation of tolerances will

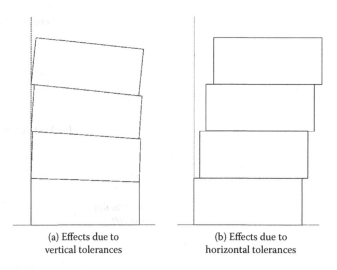

(a) Effects due to vertical tolerances

(b) Effects due to horizontal tolerances

Figure 16.3 Out-of-verticality effects of manufacturing and installation tolerances in modular construction.

be required at connections to the in situ concrete elements, such as stabilising cores and lift shafts.

Light service connections are relatively flexible, but soil and vent pipes and other large-diameter elements can require special consideration, particularly where connections are made to ground services. One way of accommodating vertical connections between large-diameter service pipes and ground works is the use of S-shaped offset bends, as the relative rotation of these can accommodate a degree of out-of-alignment.

16.2.1 Tolerances in steel-framed modules

In the British Constructional Steelwork Association (BCSA) *National Structural Steelwork Specification* (NSSS), the permitted out-of-verticality of columns in steel-framed structures is $\delta_H \leq$ height/600, but this should not exceed 5 mm per storey. Furthermore, for buildings of more than 10 storeys high, the cumulative out-of-verticality over the total building height should not exceed 50 mm.

In modular construction, there are two sources of positional error when one module is placed on another. These are due to the potential difference in the width of the modules in the manufacturing process and the practical accuracy that is possible in the positioning of the module in installation by crane.

In manufacturing, the maximum permitted tolerance in geometry of a module may be taken as illustrated in Figure 16.4. This corresponds to a permitted variation in length, width, and out-of-verticality of $h/500$, where h is the module height. This corresponds to a maximum of 6 mm for h = 3 m. The permitted bow in the side of the module between the corners is taken as $h/1000$ (or 3 mm).

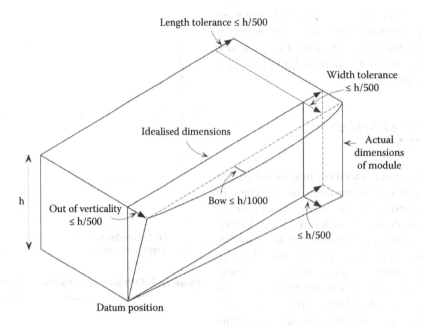

Figure 16.4 Permitted maximum geometric errors in manufacture of modules.

When considered over a large group of n modules, the average out-of-verticality of the corners of the modules may be taken as half of the permitted maximum tolerance. Therefore, the cumulative out-of-verticality in manufacture may be taken as $nh/1000$, which equates to $3n$ mm over the building height, nh.

In addition to manufacturing tolerances, the total out-of-verticality δ_H over the building height also includes the positional errors that arise due to the installation method and the form of the connections between the modules. Given the practical difficulty in placing one module on top of another module, it is proposed that the maximum horizontal out-of-alignment of the top of the one module may be up to 12 mm relative to the top of the module below. However, this positional tolerance also includes the tolerance in the manufacture of a module, which can be as high as 6 mm, as shown in Figure 16.4.

Over a group of n modules at any level in the building height, the cumulative positional error at this level due to installation of all of the modules can be partially corrected over the various levels. Lawson and Richards (2010) proposed that the maximum out-of-verticality, e, of the top of the module at any level may be taken statistically as $e < 12\sqrt{(n-1)}$ (in mm) relative to the base of the building.

For $n = 12$ storeys, this formula leads to a permitted cumulative out-of-verticality of 40 mm at the top of a modular building, and it is proposed that this maximum limit of 40 mm also applies for taller buildings. This permitted tolerance is clearly stricter than the maximum of 50 mm in steel-framed construction, even though the positional error between adjacent floor levels could be greater in modular construction.

In practice, the geometrical alignment should be measured by reference to the base by a laser line projecting from the base of the lowest module. In this way, geometric errors can be corrected progressively. These out-of-verticality tolerances are included in the design for stability of a group of modules, which is based on the notional horizontal load approach, as described in Chapter 12.

16.3 MODULE-TO-MODULE CONNECTIONS

Connections between modules are structurally important, as they strongly influence the overall structural stability and robustness of the assembly of modules. The connection between the modules is made at the top and bottom of the modules, and is often in the form of horizontal or vertical connecting plates. Access for these attachments has to be made externally to the modules, and can pose practical difficulties for certain arrangements of modules.

These connections are often made from mobile access platforms. Connections to reentrant corners of the modules can be made more easily, as shown in Figure 16.5. In this case, the end plate to the angle provides the vertical connection between the modules. Horizontal connections can be made by a bolted plate to a corner angle with a welded nut to the rear face of the angle, as shown in Figure 16.6. Some systems also include acoustic pads between the corner posts to eliminate any direct impact sound transfer, although this is not normally necessary for most building types.

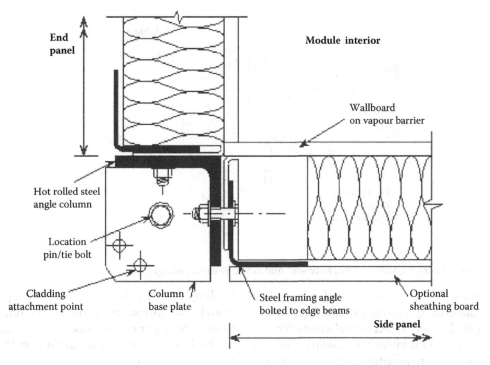

Figure 16.5 Detail of re-entrant corner of module.

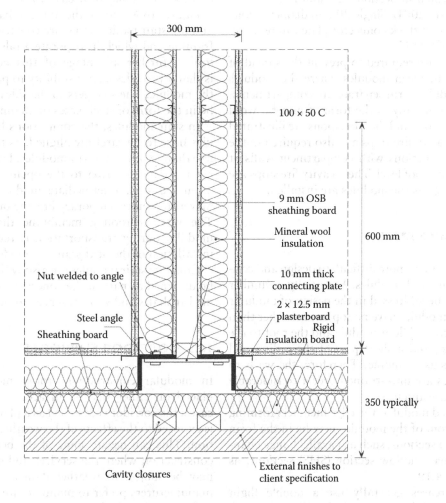

Figure 16.6 Detail at re-entrant corner of adjacent modules with connector plate.

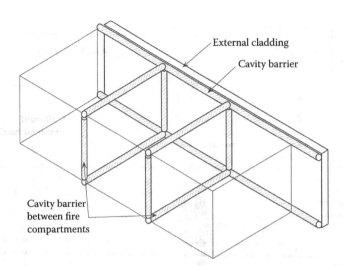

External cladding

Cavity barrier

Cavity barrier between fire compartments

Figure 16.7 Location of cavity fire stop barriers between modules to prevent passage of smoke in fire.

The connections are designed to transfer horizontal forces due to wind loading, and the extreme forces due to loss of support in the event of accidental events (known as robustness or structural integrity). Guidance on how to meet the robustness requirements for modular construction is presented in Section 12.7 and in SCI P302 (Gorgolweksi et al., 2001). Single 20 mm diameter bolts attached to a 12 mm thick connector plate can resist up to a shear force of 90 kN.

Cavity barriers are required to prevent the spread of smoke or flame between modules where the modules or groups of modules form separate fire compartments. The fire stops are usually in the form of mineral wool "socks" in wire gauze, and their locations are illustrated in Figure 16.7. Cavity fire stops are also required in the external wall at junctions with compartment walls at each floor level and roof level. These cavity fire stops are placed at each floor as the modules are installed.

16.4 MODULAR STAIRS

Modular staircases are more difficult to design and construct than room-sized modules. Essentially a number of issues have to be addressed in the design of modular staircases: Stair modules have no top or base other than the floor to the ground floor module and the roof to the top floor module, and a short part of the ceiling and floor, which acts as a landing. Therefore, the walls to staircase modules are unrestrained over a considerable length at their base and top.

For steel-framed modules, the edge members forming the top and bottom of the modules may be in the form of hot-rolled steel sections, such as parallel flange channel (PFC) or square hollow section (SHS) sections, as shown in Figure 4.19.

Modular staircases generally use a double flight of stairs supported between half landings and full

landings. The half landings generally support one flight from below and one from above. The full landing at the foot of the upper module sits on the landing at the head of the lower module. The landing at the head of the lower module is of greater depth than the one that sits on it, so that it may be used as the final step in the flight.

Figure 16.8 shows the typical base, intermediate, and top stair modules that are constructed in light steel framing and which incorporate a false landing at their top. A further advantage of this additional landing is that it provides lateral stability to part of the walls of the module. The stringers to the sides of the stairs may be in the form of steel plates or channel sections.

In stair modules, the connections between the modules have to be carefully aligned, as they are more visible than in other types of modules. During installation, a temporary covering to the open-topped modules is often required. Intermediate modules with no floors may also require temporary bracing during installation. The weatherproofing membrane that protects these modules during transportation is removed before the installation of the next staircase module.

Concrete stair modules are also widely used, and may be designed as part of the concrete core by incorporating lift shafts and vertical service zones (see Chapter 3).

16.5 CORRIDOR SUPPORTS

In modular construction, room modules are often placed on either side of an access corridor. In most cases, the corridor is designed as a planar element, but it is exposed to the effects of the weather during construction. This can negate some of the benefits of modular construction where the services and other components may be subject to weather damage. Therefore, some manufacturers prefer to manufacture modules of 12 to 15 m length, which incorporate the corridor and are

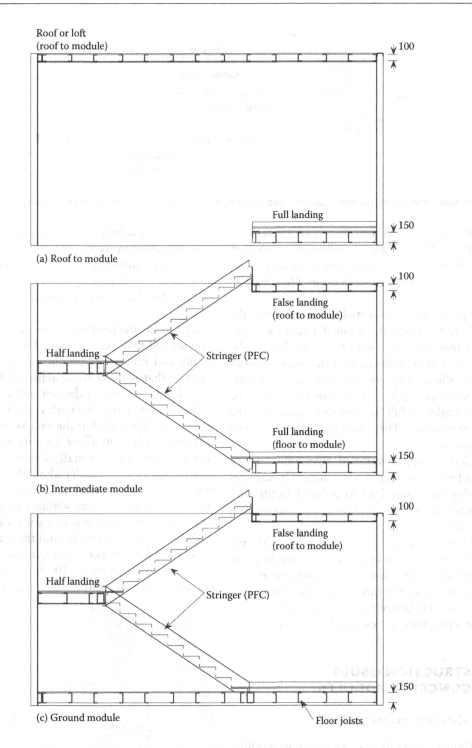

Figure 16.8 Modular stairs in light steel framing showing the double ceiling and floor at the landings.

essentially complete and weathertight during construction. However, because the corridors within the modules are open-sided, the modules may be more flexible during transportation and installation.

It is also possible to manufacture long, thin corridor modules of 6 to 10 m length between groups of two or three room modules, although this can be a complex method of construction in terms of access to the connections between the modules. This may be an option in wider corridors where double-leaf separating walls are required to the corridor for acoustic reasons, and where services can be preinstalled along the ceiling space of the corridor. The minimum module width for this form of construction is typically 2 m.

For the case where the corridors are installed in planar form, the structural depth in the corridor area is

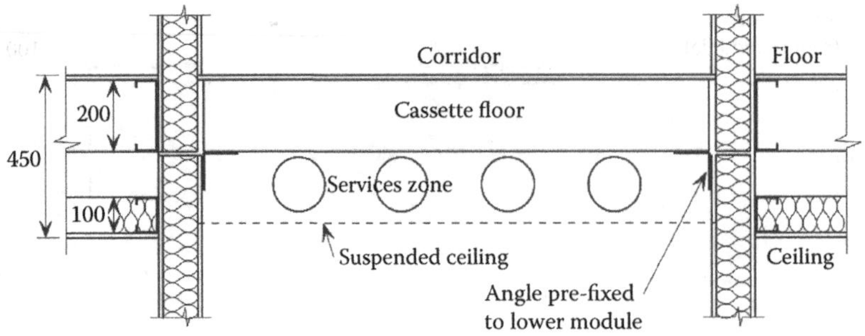

Corridor   Floor

Cassette floor

Services zone

Suspended ceiling

Ceiling

Angle pre-fixed
to lower module

200

450

100

Figure 16.9 Corridor service zones below a cassette floor system that is supported from the lower module.

shallower than the adjacent modules, as only one floor layer is required. This additional ceiling zone can then be used for the horizontal distribution of heating, ventilating, electrical, and other services, as illustrated in Figure 16.9.

The floor joists may be oriented across or along the corridor. Where the joists are oriented across the floor, the cassettes may be supported on the module at the same level or on the module below, if that makes installation easier. Where they are oriented along the corridor, the joists are supported on crossbeams (in the form of deep angles or PFC beams) that connect to the corners of the modules. These members also act as ties between modules.

A suspended ceiling is installed below the service zone and can be removed for maintenance. The services in the corridor zone may lead to reduced headroom, and in this case, the minimum floor-ceiling height in a corridor is 2.2 m.

The corridors can also be designed to provide the horizontal bracing action to open-sided modules in order to transfer wind loads to vertical bracing or concrete cores. In this case, the corridor can incorporate an additional horizontal lattice girder, or it can be braced to provide this function, as described in Chapter 12.

16.6 CONSTRUCTION ISSUES FOR CONCRETE MODULES

16.6.1 Design for construction

The installation method for precast concrete modules should be part of the initial design process, and it follows that the manufacturer should be involved early in the design process to advise on installation processes, craneage requirements, and on-site connection methods.

A method statement should be developed at the start of a project detailing how the modular or other precast concrete elements will be manufactured, transported, and installed, including details covering:

- Safety (including the mandatory safety statement)
- Handling/craneage and transportation (with appropriate consideration as to the weight of the modules)
- Site installation (procedure, sequence, location, and influence on the construction programme)

Design for the temporary conditions during installation should take into account forces in individual elements and internal joints during lifting, and also the support that is likely to be achieved from the partially completed structure in the rest of the building.

As concrete gains strength over time, the time from casting to lifting within the precast concrete factory is obviously critical to allow for a daily production cycle. The concrete mix is usually designed to enable lifting in the factory to be undertaken after 16 or 24 h. This is covered in more detail in Chapter 13. If the modular units are installed on site within a short time of production, then it is important to ensure that the concrete is sufficiently strong to withstand the forces during lifting both in the factory and on site, including any additional impact forces. Figure 16.10 shows a modular precast concrete structure during construction. The modules

Figure 16.10 Precast modules in position in a prison building. (Courtesy of PCSL.)

are provided with cast-in sockets for attachment of safety barriers around the floor while other site work is underway.

16.6.2 Connections and foundations

Connecting concrete modules and other precast elements to in situ foundations requires use of projecting starter-bars that are cast or grouted into the foundation (see Figure 16.11). The precast module or element can then be craned onto the seating area of the foundation, and the starter-bars are inserted into holes in the precast unit. The module is placed on steel shims that are prelevelled on the correct line. These joints are then grouted full after final alignment.

The seating area on the foundation that receives the modules should be designed with adequate tolerances and should not be in a restricted position. The foundation seating area should be designed for possible impact loads during installation as well as the self-weight of the module.

16.6.3 Screeds

Levelling screeds are used in areas where the modules connect to one other, such as in corridors and entrance ways to each unit. Within the module, the use of a screed should be avoided if possible by finishing the concrete to receive the floor covering. Further details on the use of screeds in concrete modules are given in Chapter 15.

16.6.4 Balconies

Balcony units in concrete modules are manufactured with steel reinforcement projecting from the back,

Figure 16.11 Installation of open-sided modular precast concrete units onto ground beams. (Courtesy of Oldcastle Precast.)

which can be connected to the steel reinforcement in the main module. This is often carried out by threading the reinforcement into a pre-prepared hole, which is later grouted up. The balcony units are temporarily supported until the grout or concrete and any finishing materials have been placed and the concrete has gained sufficient strength.

16.7 TRANSPORT OF MODULES

Transportation requirements for wide loads along motorways and trunk roads in the UK are summarised in Table 16.1. The transitions in external module widths for different transportation requirements occur at 2.9, 3.5, and 4.3 m, depending on the type of vehicle. These correspond to internal module widths of approximately 2.6, 3.2, and 4 m.

When the module is more than 3.5 m wide, police notice is required along the delivery route from the manufacturer to the site. However, if the site is within 300 km travel distance from the delivery point, then it is possible to have only one driver and no mate in the vehicle for modules up to this width. Often for logistical reasons, the delivery lorries are required to wait for a short time at suitable locations (normally a motorway service station or another holding point) until they can be received at the correct time and order on site.

Generally for design in modular construction, a module width of 4.3 m should be considered to be the sensible maximum for transport on the trunk road network. However, loads up to 5 m wide can be transported by an articulated lorry with police escort. Two smaller modules up to 2.9 m width and 7 m length may be placed on one lorry to minimise transport costs, as shown in Figure 16.12. This may be the case for relatively narrow student study bedrooms and hotel rooms.

Table 16.1 Summary of UK law requirements for width of transport vehicles

Type of vehicle	Width of load	Vehicle mate required	Police notice required	Other notice
Construction and use (C&U)	≤2.9m			
Special type	≤3.5m		✓	
Both C&U and special type	≤4.3m	✓	✓	
Indivisible load on C&U vehicle	≤5m	✓	✓	Form VR1

Source: Department of Transport, The Road Vehicles (Construction and Use) Regulations, 1986, www.legislation.gov.uk.

Note: Additional width requirements on local roads may apply.

Figure 16.12 Delivery of two modules on a lorry. (Courtesy of Unite Modular Solutions.)

Further restrictions may apply for railway bridges, where the maximum load height is generally 4 m. A low loader vehicle should be used if the module height exceeds 2.8 m. Road widths and turning circles in suburban areas may limit the type of vehicle and width of load that may be used. Access should be investigated early in the project, as it may influence the size of modules that may be used, and their means of installation. Limits may also be placed on the axle weight and total weight of the vehicles passing over some minor bridges and over culverts, etc.

Road closures may be required for narrow access roads, and therefore the installation should be linked to either quieter periods of use, e.g., 10 a.m. to 4 p.m., or certain days of the week. Residents may have to be informed about roadside car parking on days where installation is planned. An installation rate of up to 10 modules per day can be achieved in the summer months, but a sensible installation rate should be agreed on for the project, taking account of site difficulties, weather and light conditions, etc.

16.8 CRANEAGE AND INSTALLATION

16.8.1 Lifting of modules

Modules are generally lifted from their corners using a lifting beam or a frame that minimises the inward component of force exerted on the module due to the force in the inclined cables. Some lightweight modules or pods are lifted from their base to avoid damage to the internal finishes.

The various forms of lifting systems are illustrated in Figures 16.13 to 16.16, and are described as follows:

- Inclined cables to the corners of the modules (only possible for modules with corners posts and heavy edge beams at ceiling level) (see Figure 16.13)

- Single lifting beam with four inclined cables to the corners of the module
- Main lifting beam and crossbeams with vertical cables to the corners of the module (see Figure 16.14)
- Rectangular lifting frame (often composed of welded steel hollow sections) that permits lifting from intermediate points on the frame (see Figure 16.15)
- Lifting frame manufactured with a protective cage for unhooking the modules by an operative attached to a lanyard (see Figure 16.16)

Sufficient space on the site is required for the delivery lorry, as the modules are usually lifted directly from the lorry into position. Therefore, deliveries have to be carefully timed to avoid congestion at the site. Suitably sized cranes are required to install the modules, and for light steel modules, a 100-tonne mobile crane is often required when the lifting boom is extended to its maximum distance (around 25 m). Tower cranes are often used in high-rise construction but generally cannot lift heavy loads at their full extension. Therefore, the size and positioning of cranes to install the modules require detailed site planning.

Various issues should be taken into consideration when choosing a suitable crane and the module installation sequence, including

- On-site and public safety
- Access for the mobile crane
- Module dimensions and weights
- Maximum reach of the crane to the module location
- Site constraints, such as overhead power lines
- Ground-bearing pressures for the crane legs

For all types of modules, an additional force of 25% more than the self-weight of the module should be

Figure 16.13 Lifting of module from the corner posts without a lifting beam. (Courtesy of Yorkon.)

Figure 16.14 Lifting of light steel module by main beam and crossbeams. (Courtesy of Futureform.)

considered to take account of dynamic forces during lifting (in addition to normal factors of safety). All lifting beams, shackles, and cables should be load tested to an overall factor of safety of at least two before being used, and they should be load tested regularly.

16.8.2 Lifting points in steel modules

The forces developed in the modules during installation can be higher than in service. Therefore, it is important that the implications of the method of lifting are taken into consideration in the design of the modules.

Figure 16.15 Lifting of light steel module from a rectangular frame. (Courtesy of Open House AB.)

Figure 16.16 Lifting of module using a protective cage attached to the crane hook. (Courtesy of Ayrshire Framing.)

Figure 16.17 shows the various methods of lifting the modules, some of which might lead to high horizontal forces. Lifting points in light steel modules are normally at their corners, particularly if shackles or cables can be attached to the corner posts.

The weight of a fitted-out light steel module can be 7 to 12 tonnes (equivalent to 3 to 4 kN/m² of floor area), increasing to 15 to 25 tonnes for a light steel module with a concrete floor. The preferred method of lifting of heavy weight modules is by a two-dimensional frame, so that the forces acting on the modules are vertical. Temporary bracing will often be required in open-sided modules.

16.8.3 Lifting points in concrete modules

The lifting points on a precast concrete module are cast into the unit during production and are designed where possible so that the out of plane forces are minimised in the lifting operation (see Figure 16.18). Four-point or sometimes eight-point lifting will generally be required for concrete modules, depending on their size, so that loads are applied vertically (as in Figure 16.19). For modules with concrete ceilings, inclined forces can be applied, as shown in Figure 16.20, but this is not generally possible for open-topped modules.

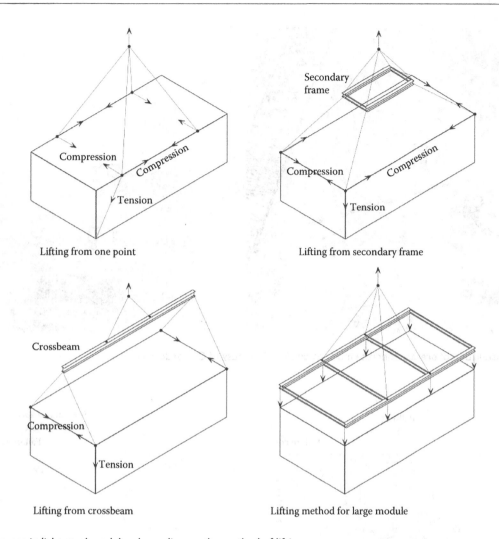

Figure 16.17 Forces in light steel modules depending on the method of lifting.

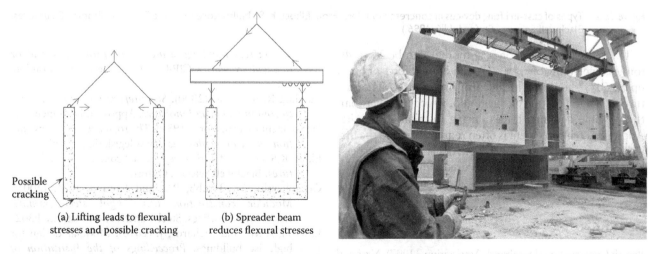

Figure 16.18 Lifting precast modules (a) incorrectly, resulting in flexural cracking, and (b) correctly using a lifting beam. (From Elliott, K.S., Multistorey Precast Concrete Framed Structures, Blackwell Science, Oxford, UK, 1996.)

Figure 16.19 Lifting a concrete module using a spreader beam. (Courtesy of PCSL.)

Figure 16.20 Installation of precast concrete modules on site. (Courtesy of Oldcastle Precast.)

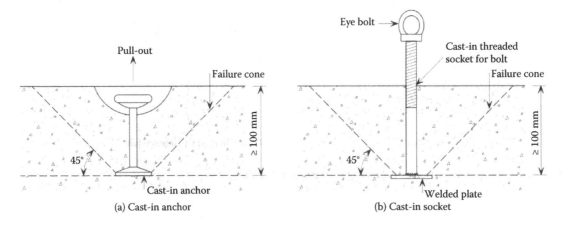

Figure 16.21 Types of cast-in lifting devices in concrete modules. (From Elliott, K.S., *Multi-storey Precast Concrete Framed Structures*, Blackwell Science, Oxford, UK, 1996.)

The self-weight of a concrete module can be 25 to 40 tonnes, depending on its size. An additional force of often up to 50% should be included in lifting after manufacture due to the suction of the mould, which is higher than the possible dynamic forces during installation on site. Additional reinforcement is often required around the lifting points to prevent cracking, particularly near corners. Proprietary devices can also be used, which reduces the need for additional steelwork (as shown in Figure 16.21).

REFERENCES

British Constructional Steelwork Association. (2007). *National structural steelwork specification for building construction.* 5th ed.

Brooker, O., and Hennessy, R. (2008). *Residential cellular concrete buildings: A guide for the design and specification of concrete buildings using tunnel form, crosswall or twinwall systems.* CCIP-032. Concrete Centre, London, UK.

Building Regulations. (2000). *Site preparation and resistance to contaminants and moisture.* Approved Document C.

Department of Transport. (1986). *The road vehicles (construction and use) regulations.* www.legislation.gov.uk.

Elliot, K.S. (1996). *Multi-storey precast concrete framed structures.* Blackwell Science, Oxford.

Gorgolewski, M., Grubb, P.J., and Lawson, R.M. (2001). *Modular construction using light steel framing: Residential buildings.* Steel Construction Institute P302.

Lawson, R.M., and Richards, J. (2010). Modular design for high-rise buildings. *Proceedings of the Institution of Civil Engineers: Structures and Buildings*, 163(SB3), 151–164.

Chapter 17

Factory production of modules

This chapter reviews the manufacturing processes for steel, timber, and concrete modules, which depend in part on the potential variation of the output in terms of geometry, structure, and finishes. Steel- and timber-framed modules can be manufactured on various forms of semi-automated production lines. The production of concrete modules should take account of the additional manufacture of moulds or formwork, and also the turn-around in casting and lifting.

17.1 BENEFITS OF OFF-SITE MANUFACTURING

Manufacturing of modular units in a factory environment can take many forms, from the simplest replication of site-based construction to a sophisticated production line manufacture similar to those used in the automobile industry. The economics of the manufacturing process are influenced by how flexible the modular system is required to be at the design stage. In principle, the discipline of modular construction requires that some degree of standardisation is achieved, which leads to economies in manufacturing and in the materials procurement processes.

There are many potential benefits of factory production compared with traditional site-based techniques. These include the following:

- Ability to achieve a rapid, reliable construction programme by reduced exposure to risks, such as availability of site trades and adverse weather conditions
- Simplified (and unified) procurement routes
- Reduced waste and damage to materials, components, and finishes on site
- Efficient preordering, delivery, and storage of materials in factory conditions
- Increased productivity in the factory and also on site
- Mechanisation of the manufacturing process, including overhead lifting, use of sophisticated machine tools, etc.
- High level of quality control, thereby avoiding reworking and delays

- Dry construction process in the factory and in on-site work
- More reliable installation of sensitive services and equipment, for example, in medical buildings
- Ability to manufacture modules for remote sites or use in severe environments, where site construction would be otherwise expensive or logistically difficult
- Economy of scale of manufacture with minimal downtime in manufacturing
- Production rates can be matched to site delivery, or the modules can be stored for a short period in a yard next to the factory
- Improved health and safety by reducing risks in a controlled production environment
- Highly skilled and well-trained workforce in factory-based production with continuity of employment

These benefits are independent of the materials used in the modules, but the choice of material will influence the manufacturing and construction process.

17.2 MANUFACTURE OF LIGHT STEEL AND TIMBER MODULES

Various forms of factory production of light steel and timber modules exist with their different levels of automation and economic models. Mullen (2011) describes the manufacturing processes for timber framing in the United States, where sophisticated factory production of about 85 standardised house types has been achieved across many states. However, he notes that even in the most productive factory, the average production is about 65% of peak capacity over the year, when allowing for setting up and downtime between production runs. Senghore et al. (2004) and Mehotra et al. (2005) also reviewed the optimum layouts of modular housing factories using timber framing.

High levels of automation increase productivity in the factory operations, but the investment has to be recouped over a large production output. Furthermore, automation can also constrain the range of modules that can be manufactured, as production facilities tend

Figure 17.1 Pre-manufacture of steel floor and ceilings in hot-rolled steel and light steel elements. (Courtesy of Caledonian Modular.)

to be designed to accommodate particular maximum panel sizes and buildups of the elements (e.g., wall linings and other details).

In choosing the optimum production method, a balance has to be achieved between improved productivity and the capital investment required versus design flexibility to meet the market demands. The optimum balance will usually reflect the target markets of the particular modular manufacturer. Varied and more flexible modular designs are more readily achieved by static production or less automated forms of linear production. Manufacturers targeting high-volume repetitive buildings, such as hotels and student residences, tend to use more advanced production systems based on various levels of automation.

The three main forms of manufacturing systems for modular units are here termed static, linear, and semi-automated linear production.

17.2.1 Static production

Static production means that the module is manufactured in one position, and materials, services, and personnel are brought to the module. In the case of a steel module comprising steel corner posts and edge beams, these linear steel components are fabricated in a separate location and are assembled as the first stage in the process. Similarly, light steel walls and floor and ceiling panels may be manufactured separately or off-line. Typical stages in static production are illustrated in Figures 17.1 and 17.2. The geometry of the modules has to be precisely controlled by manual methods, although a steel-framed jig is often used to control the accuracy of the vertical placement of the walls.

The space around the modules must be sufficient to provide for temporary storage of materials and prefinished components, such as windows, which are lifted into place by hand or by using an overhead crane. The rate of construction is controlled by the availability of personnel for the specialist tasks at the required time. Therefore, the process can be relatively slow, but conversely, the critical path is not the completion of any one task. When completed, the module is lifted out by overhead crane and then either stored temporarily or transported to site.

Typically in a large factory space, up to 30 modules may be under construction at any one time. The cycle from assembly to completion of a module is typically 3 to 7 days, allowing for painting and drying time, etc. This arrangement suggests an average production rate of 4 to 6 modules per day, and an output of 800 to 1200 modules per year may be achieved with modest downtime between orders.

In static production, the production and access space should be on average four to five times the module size, which is equivalent to about 5000 m² of factory space for production of 20 to 30 modules at a time. The height of the overhead crane should be at least 8 m to allow for clearance of the module and its lifting frame, and lifting of one module over another in order to remove completed units from the factory floor into temporary storage or delivery to the site.

17.2.2 Linear production

Linear production means that the manufacturing process is sequential, and is carried out in a discrete number of individual stages that is analogous to automotive

Figure 17.2 Static production of modular units. (Courtesy of Caledonian Modular.)

Figure 17.3 Assembly of light steel panels in production. (Courtesy of Unite Modular Solutions.)

production lines. Geometric accuracy of the panels is achieved by use of planar jigs, as shown in Figure 17.3. Individual panels are boarded with plasterboard, and sheathing boards fixed using specialist fastenings, such as air-driven pins.

The modules may be manufactured on fixed rails or trolleys and moved between stations. Each station has a number of production teams or trades associated with it and a prescribed zone on the factory floor. The key

difference between this form of production and static production is that the modules are moved between dedicated stations, rather than the production teams having to move from module to module.

The number of production stages is dependent on space and production volumes, but generally each will reflect well-defined operations, such as plasterboarding, installation of bathrooms, decorating, etc. In designing the facility, the time involved in each stage should be similar to avoid bottlenecks, and to balance the "dwell time" at each stage on the line. The design of modules should also reflect the sequential nature of the line, such that all of the tasks associated with any particular station can be discharged in a single operation. Modules are typically moved by electrical or similar vehicles, motorised trolleys, or manually on roller tracks, where their size permits.

Materials and components are stored in bulk next to the stage where they are used. These are often delivered via a wide access route parallel to the production line. Work teams can be supplemented if production slows, and use of contract labour for less specialised tasks is relatively common. Often two, three, or four staged production lines operate in parallel and can share the same materials handling and storage areas. The factory space is by its nature relatively long (typically 60 to 100 m), and a 12 to 15 m wide zone is allowed for each production line (excluding the additional materials storage areas).

Output rates reflect the size and sophistication of the line, but a four-line facility producing 3 modules per line per day equates to a maximum output of up to 3000 modules per year. The rate of production can be matched to the delivery to the site, but temporary external or internal storage may be required to maximise throughput and to meet peak demands.

Late design changes often result in disruption to the manufacturing process, which can lead to delays while jigs are changed or materials ordered. Also, production lines rarely operate at full efficiency when using new components or following major product modifications. Changes after design "freeze" or "sign-off" should therefore be avoided, as they may incur cost penalties. However, manufacture of the modules can be carried out in parallel with on-site activities, and modules are generally delivered "just in time" to the site.

17.2.3 Semi-automated linear production

Modern semi-automated factories for modular production are based on the same principles of conventional linear production as non-automated lines, but tend to have more dedicated stages. Typically, automated facilities have separate lines for manufacture of wall, ceiling, and floor panels as light steel-based facilities often incorporate on-line roll-forming machines for each type of panel.

Automated lines commonly include facilities for creating window and door openings (often by the incorporation of subassemblies), and installing insulation and inbuilt services, such as cabling and telecoms. They do not normally include automated systems for bathroom fit-out and installation of fitted furnishings. Bathroom pods are often prefabricated off-line or brought in. Furnishings are generally more difficult to automate and

so become follow-on operations. Semi-automated lines therefore tend to comprise a highly automated series of operations requiring specialised equipment, followed by a series of relatively conventional manual operations.

The speed of manufacture of panel lines has to be matched against the more manual operations at the end of the production cycle. Panel manufacture can be optimised by use of turning or "butterfly" tables that permit panels to be worked on from both sides. Special stages may include spray painting and forced drying.

These facilities have been designed for production rates of one module every 20 min or up to 30 modules per day. This corresponds to an output capacity of approximately 6000 units per year. However, production demand tends to be cyclical and even seasonal for sectors, such as student residences. Actual expectancies for output therefore tend to be lower, at perhaps 50 to 65% of the theoretical capacity. Typically, a typical factory space of 10,000 m² should be allowed for a semiautomated production line, materials storage, and office space.

The rate of production often exceeds the rate of delivery to the construction project, and so temporary storage of the modules is required. Where there is sufficient storage space indoors, the modules are protected in the normal manner, but where storage is outdoors, special protection is required to prevent wind-driven rain penetration and the deleterious effects of moist ambient air. Modules are often stacked temporarily in groups two or three high, and they are lifted by mobile cranes onto lorries in the order required for a given project.

17.3 AUTOMATION IN PANEL PRODUCTION

The capital cost of factory production depends on the degree of mechanisation and automation that is provided. Balanced against these fixed capital costs are savings due to more efficient production technologies, more efficient ordering and use of materials, higher-quality levels, and time-related savings. These economics are examined in Chapter 18. A semi-automated production line is claimed to be three times more productive in terms of the factory-based operations than the equivalent on-site construction.

The expansion of modular construction has in part been facilitated by the availability of sophisticated numerically controlled machinery and integrated CAD/CAM software. Designs developed on computer-aided design (CAD) systems can be intelligently interpreted to generate computer-aided manufacturing (CAM) production data.

Many modern production facilities dealing with either timber framing or light steel framing have highly automated frame assembly systems. While these vary in design, they typically comprise a series of stations, each

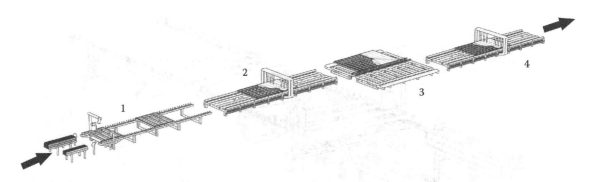

Figure 17.4 Example of semi-automated line for wall and ceiling panels. 1 = framing station, 2 and 4 = working tables with multifunctional bridge, 3 = turning table.

dedicated to particular assembly operations arranged sequentially in the form of a production line.

In the case of timber, the stock is cut sections, and in the case of steel, sections are normally rolled in the factory from steel strip of the correct width, or lengths of cold-formed sections are brought in from outside manufacturers. Small-packaged cold-rolling machines are commonly used in modular production. The CAD/CAM machinery ensures that the sections are cut to the correct length and prepunched for screw holes and services for easy assembly. Panels may weigh 60 to 200 kg in lengths of 5 to 10 m. Panel throughput will vary, but may be around 20 to 30 m per hour, depending on the roll-forming machinery.

Figure 17.4 illustrates one typical form of a modern production facility into which subassemblies, insulation, and services might be fed using ancillary production equipment. Most product lines centre around three key elements: framing stations, working tables,

and turning tables. The operation of these stages is described as follows.

17.3.1 Framing stations

Framing stations used to manufacture wall, ceiling, and floor panels are generally manual or semi-automatic. Manual stations require C section studs and "tracks" to be introduced by hand (see Figures 17.3 and 17.5). Often the top and bottom tracks will be premarked and moved through the station, and the wall C sections are introduced as they progress. In automated systems (see Figure 17.6), C sections will be located automatically from a feeder machine. The panel progresses incrementally through the station and the C sections are positioned and secured. Connections between the C sections are typically nailed or riveted for each panel in its static location.

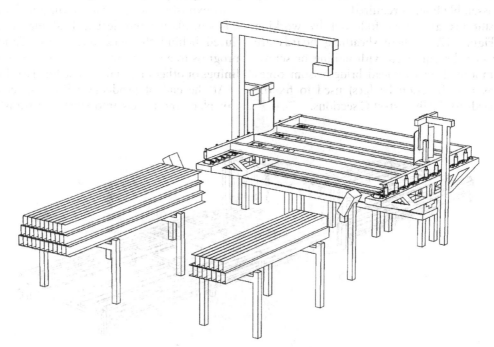

Figure 17.5 Manual framing station for light steel framing panels.

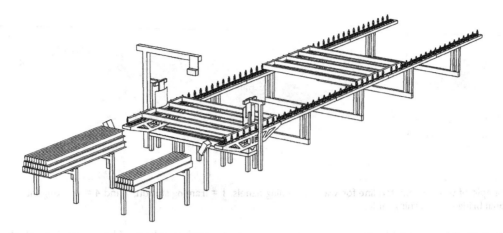

Figure 17.6 Semi-automated framing station.

This basic approach is ideally suited to loose timber framing where studs can simply be dropped between rails. In light steel-framed panels, the C sections are nested into the open web of the top and bottom tracks. Commonly, as for timber, light steel panels move relative to a stationary fixing bridge. Automated systems are particularly appropriate for the production of large panels, typically up to 10 m long.

Other types of machines comprise simple flat tables where manually or automatically placed C section studs are set out and nailed (sometimes in conjunction with sheathing boards), but where the workpiece is stationary.

Where windows, doors, or other openings are required, subassemblies can be incorporated to frame-out openings. These are often produced off-line in dedicated production areas and transported automatically to the main assembly line, as required.

Framing stations are often followed by working tables (see Figure 17.7) where sheathing boards are fixed. These may be equipped with nailing or screwing devices mounted on overhead bridges (sometimes referred to as multifunction bridges) used to fix the sheathing boards to the light steel C sections.

Where window or door openings are required in the sheathing boards, the studwork or C sections have to be arranged to accommodate these (either by altering the spacing of the C sections locally or by the introduction of subassemblies). Saws or router devices on the bridge can cut the appropriate aperture in the sheathing boards.

17.3.2 Turning tables

Turning tables generally comprise two adjacent tables that rotate and raise panels in a vertical arc about one long edge while supported on one table, and then gently tip them onto an adjacent table that is also approximately vertical at this point. The second table then lowers the panel such that the face that was formerly beneath is now on top. This process is illustrated in Figure 17.8.

By exposing the open face of the panel, turning tables allow insulation and electrical cabling, etc., to be introduced behind the sheathing board. Panels can then progress to a second working table where plasterboard linings or other materials can be fixed to close the panel.

At the end of production lines, special machinery can place the panels into stacks or transfer panels to

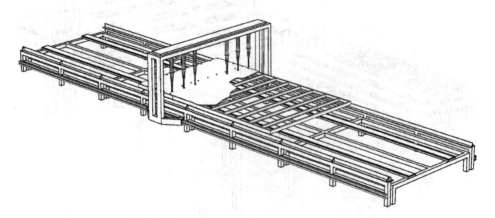

Figure 17.7 Working table with nailing bridge for sheathing boards.

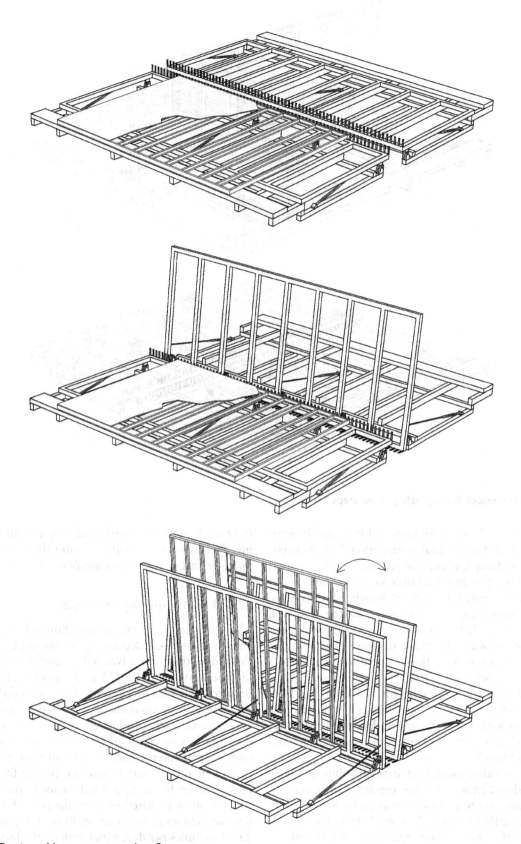

Figure 17.8 Turning table process steps 1 to 5.

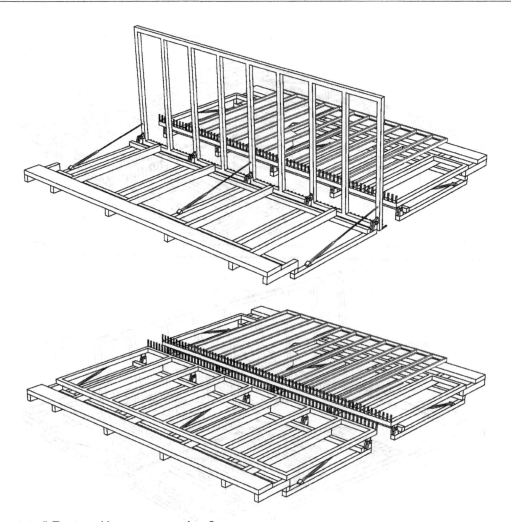

Figure 17.8 (continued) Turning table process steps 1 to 5.

assembly areas. In some systems, a lift bridge is automatically guided to the final workstation. This incorporates lifting clamps, which are engaged to hoist panels either singularly or in stacks over to a roller bed.

From this point, the stack of panels is transferred to an assembly area, where the floor, ceiling, and wall panels are assembled to form a three-dimensional module. In systems where there are separate lines for wall, ceiling, and roof panels, this is the point where lines will typically combine.

Panel identification is important and identifiers have to be accessible. Codes or bar codes are often used in conjunction with a quality control document that catalogues such aspects as inspection requirements and inspection history.

Panels are often assembled manually into modules using overhead cranes or other specialist lifting equipment. Panels may be connected using a variety of methods, but usually fixings will be made from the outside (usually in the case of light steel using bolted connections or self-tapping screws and brackets) so that the internal linings are unaffected. Sophisticated measuring devices may be used to ensure the squareness and good overall geometrical control of modules.

17.3.3 Finishing operations

The modules are serviced and finished in a series of later workstations depending on the sophistication of the production. For advanced production of 10+ modules per day, a total of 25 to 30 workstations may be required, whereas for outputs of 2 to 4 modules per day, the finishing operations require fewer workstations, and indeed, static production may be preferred. Often painting is a mechanised task, but first fix servicing is more labour-intensive and can require longer times.

Mullen (2011) states that for timber-framed modules used in housing, a total factory space of about 14,000 m^2 is required or a production of 10 modules per day, reducing to about 5000 m^2 for production of 2 to 4 modules per day, which reflects the larger number of workstations that are required for the higher output.

Modules are rarely stored for more than few days in the factory when completed, and so they have to be stored temporarily until required on site.

17.4 FIELD FACTORIES

For large projects, it may be possible to set up field factories in warehouses or similar large plan buildings close to the project. This generally takes the form of static production if there is a suitable overhead crane. The field factories processes may use normal building techniques manned by trade operatives.

This facility should be close to the project (say within 10 miles) and should have good access for delivery of the modules and materials. The optimum minimum size of a project for a field factory tends to be in the range of 200 to 400 modules, which allows for rental of the space for a period of perhaps 6 months. A production rate of two to three finished modules per day would be expected from this temporary factory. Premanufacture and temporary storage is required as average installation rates on site may be expected to be twice the production rate.

17.5 MANUFACTURE OF PRECAST CONCRETE MODULES

17.5.1 Manufacturing process

Concrete modules are cast in moulds or in formwork with their reinforcement fixed in place. They are often manufactured with an open base so that the walls and ceiling can be cast monolithically, as shown in Figure 17.9. The ceiling then forms the floor to the module above. The

Figure 17.9 Casting a modular unit from the top using self-compacting concrete. (Courtesy of Tarmac Precast Ltd.)

walls of the modules may be recessed to support planar floor slabs, such as at corridors.

After placement, the concrete on the top surface is smoothed and covered while it cures. The concrete will usually gain sufficient early-age strength overnight, so that it can be lifted out of the moulds into storage. The top surface may require further "power floating" to achieve the required smoothness.

17.5.2 Design coordination

Lead-in times between order and delivery of modules are important whatever the material, but bespoke or complex precast concrete modules can take longer to produce, as they require additional time for detailed design, and special mould or formwork production.

Drawings produced by the precast module manufacturer show the position of fixings, penetrations, cast-in items, openings and lifting anchors, and the location and size of service voids. However, before the design is finalised, coordination with the wider project team is vital, in particular for the services, cladding, and foundations.

17.5.3 Use of self-compacting concrete (SCC)

Self-compacting concrete (SCC) possesses superior flowability without segregation, thus allowing self-compaction. If designed and placed correctly, it is also able to provide a more consistent and superior finished product with fewer defects. Another advantage of using SCC is that less labour is required in order for it to be placed and made good after casting.

The majority of precast concrete plants in the UK, Europe, and United States have now converted to SCC, although in conventional in situ construction, SCC is less often used. Precast concrete manufacturers have their own on-site batching plant, and so they are able to take full advantage of all the benefits of SCC, such as higher quality, faster placement, and less labour.

A standard SCC mix specification typically achieves a cube strength of 26 to 28 N/mm^2 after less than 24 h. After 9 days, the required 50 N/mm^2 design strength is generally achieved.

17.5.4 Casting process

Before casting, reinforcement is fixed into the moulds, and it must be securely fixed, as the force of placement of the fresh concrete can displace the reinforcing bars. Other fixtures, such as pipes, electrical wiring and conduits, fixing sockets (for internal attachments), box-outs, and window and door frames, must also be

Figure 17.10 Precast modular prison units being moved to storage. (Courtesy of Precast Cellular Structures Ltd.)

accurately secured at this stage. Lifting attachments are often cast into the modules to enable them to be lifted out of the moulds by crane, usually the morning after casting.

For production of standard precast elements, concrete is poured into cleaned and oiled steel moulds, which are dimensionally accurate to a maximum of ±3 mm (see Figure 17.10). Architectural cladding is often cast in timber or fibre-glass moulds in order to create the required surface finish. Vibrators may be clamped to the formwork, which are tuned to the correct oscillations for the size and weight of the filled mould to ensure an even compaction of the concrete in the mould. Handheld vibrators are sometimes used to ensure the concrete fills all areas of the formwork, particularly at corners.

17.5.5 Concrete finishes

High-quality finishes can generally be achieved with precast concrete, due to a combination of high-quality formwork, use of SCC, and consistent workmanship in a controlled internal environment (Figure 17.11). Precast concrete should achieve a type B finish according to BS 8110 (1997). If required, a type C finish can also be achieved, but this is likely to be more expensive.

An alternative to specifying a type C finish is to use a suitable paint or skim coat of plaster. Other systems are available, such as fillers, which can be used instead of gypsum plaster, and which can prove to be more cost-effective. If a quality finish is required on both sides of the wall, then it should be cast vertically so that both faces are cast against formwork.

Precast concrete cladding is available in a wide variety of low-maintenance and durable architectural finishes, including self-finished options and a range of applied materials, as shown in Figure 17.12. However, due to the monolithic casting of the modular precast concrete units, the range of suitable finishes is much less than in other precast units.

(a)

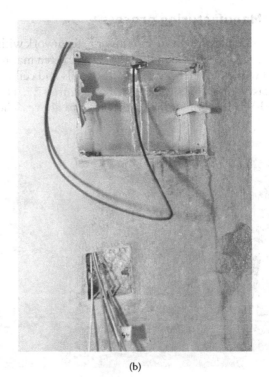

(b)

Figure 17.11 Electrical services cast into a precast concrete unit (a) placed in the top slab before casting and (b) wiring installed after casting and striking of the mould. (Courtesy of Tarmac.)

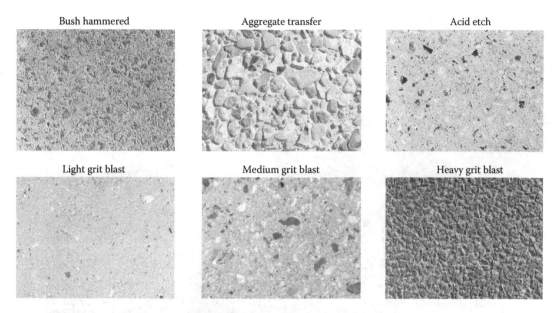

Bush hammered   Aggregate transfer   Acid etch

Light grit blast   Medium grit blast   Heavy grit blast

Figure 17.12 Examples of architectural finishes to exposed concrete. (Courtesy of Concrete Centre, *Precast Concrete in Civil Engineering*, Report TCC/03/31, London, UK, 2007.)

Modular precast units are often left with an exposed cast finish, but they may be finished on their external face with bonded brick-slips, tiles, or stone facings, such as granite, limestone, and slate. Stone finishes are often 30 to 50 mm thick. This type of finish is often used in hotels and speeds up the construction time dramatically.

Internally, the quality of finish is sufficiently smooth to accept directly applied paint or paper finishes. For less smooth surfaces, a skim coat of plaster is required. The exposed concrete surface is also very durable and hard wearing, and provides a high level of security for separating walls in secure or high-value accommodations.

17.5.6 Screeds

Levelling screeds are often used with precast concrete modules, but usually only in areas where the modules connect to corridors and in the entrances to each unit. The use of a screed should be avoided if possible by finishing the concrete in the module floor so that it is suitable to receive the final flooring or wall covering. Screeds are more common when bathroom pods are installed in order to bring the general floor level up to that of the bathroom floor. The minimum thickness of a bonded levelling screed should be 25 mm. When specifying a screed, other criteria should also be considered, such as drying time, slip, abrasion, and impact resistance, type of traffic on the floor, appearance and maintenance, and type of floor covering to be applied.

17.6 WEATHER PROTECTION

Modules are weather protected by a protective "breathable" shroud that prevents water ingress but allows for vapour to pass. A good example is shown in Figure 17.13. It provides protection during storage, transport, and installation. When installed, the shroud is cut back and taped at doors and windows, but is left in place.

Gaps may be left in the shroud for the lifting brackets, or the brackets can be bolted through the shroud. Modules are lifted by a steel frame that is load tested regularly and is rated to at least twice the finished weight of the module. Guidance on lifting systems is presented in Chapter 16.

Single modules are transported on a low trailer, and in some cases, two modules may be delivered subject to width and length limitations and access/turning constraints.

Figure 17.13 Protective breathable shroud around a module for delivery to the site. (Courtesy of Unite Module Solutions.)

REFERENCES

British Standards Institution. (1997). *Structural use of concrete. Part 1. Code of practice for design and construction.* BS 8110.

Concrete Centre. (2007). *Precast concrete in civil engineering.* Report TCC/03/31, London, UK.

Mehotra, M., Syal, M.G., and Hastak, M. (2005). Manufactured housing production layout design. *Journal of Architectural Engineering*, American Society of Civil Engineers, vol. 11, no. 1, 25–34.

Mullen, M.A. (2011). *Factory design for modular home building.* Constructability Press, Winter Park, FL.

Senghore, M., Hastak, T., Abdelhamid, S., AbuHammad, A., and Syal, M.G. (2004). Production process for manufactured housing. *Journal of Construction Engineering and Management*, American Society of Civil Engineers, vol. 30, no. 5, 708–718.

Chapter 18

Economics of modular construction

The economic principles that underline the use of modular construction are presented in this chapter. The primary economic benefit is the speed of the construction process, but the economic value of early completion depends on the particular business operation and the potential for reduced cash flow and additional revenue. This can be readily quantified for a hotel chain or a time-constrained operation, such as a university or school, but may be less apparent for a private house builder selling into a speculative market.

Buildoffsite (2009) has presented a useful guide on the procurement process, *Your Guide to Specifying Modular Buildings: Maximising Value and Minimising Risk*, which is relevant to this discussion.

18.1 ON-SITE CONSTRUCTION VERSUS OFF-SITE MANUFACTURE

Davis Langdon and Everest (2004) identified the important factors in a cost model of off-site manufacture in which the additional costs of a permanent manufacturing facility and dedicated workforce have to be balanced against tangible savings in the slow, inefficient, and wasteful site processes in more conventional construction. The economic benefits of off-site manufacture of modular systems arise from

- Economy of scale in manufacture (dependent on the production volume)
- Reduced material use, and less wastage and disposal costs
- Higher productivity in manufacture and less work on site, leading to savings in labour costs per unit completed floor area
- Higher quality and hence reduced "snagging" or rework costs
- Savings in site infrastructure and management of the construction process (known as site preliminaries)
- Savings in external consultant fees, as most of the detailed design is provided by the modular supplier
- Financial benefits to the client and main contractor resulting from speed of completion on site

Most modular construction projects involve a proportion of site work (30 to 50% being typical of the value of the whole project that is completed on site). The broad cost breakdown of a similar multistorey residential project in either site-intensive construction or modular construction is presented in Figure 18.1. In interpreting this cost breakdown, some observations from recent project are as follows:

- Materials use and waste are reduced because off-site manufacturing processes lead to more efficient bulk ordering of materials in the correct sizes for the particular project, and to less site damage. This can lead to a significant reduction in materials use by up to 20%.
- The total number of site personnel is reduced to about half of those required for on-site construction (see later), and the site personnel is mainly required for foundation work, cladding, and servicing of the nonmodular parts of the building.
- Factory personnel, materials, and overhead costs in the ex-works cost of a module often amount to 50 to 60% of the value of the completed building, of which the factory running costs can represent 30% of the ex-works cost of a module.
- Transport and other equipment costs on site are greatly reduced even though the modules are lifted by crane, because the many deliveries of materials and equipment costs are minimised.
- Site overhead and management costs (preliminaries) are reduced at least in proportion to the overall construction programme. Site preliminaries would be expected to reduce from typically 15% of the build cost to 7 to 8% in modular construction projects.

Mullen (2011) in his book *Factory Design for Modular Home Building* presents information on the breakdown of costs of timber-framed modules used in single-family houses in the United States. He states that 45 to 50% of the cost is in materials, 35 to 45% in factory overheads, and an average of only 16% in labour, which reflects the highly mechanised nature of this type of standardised modular construction. Relatively little

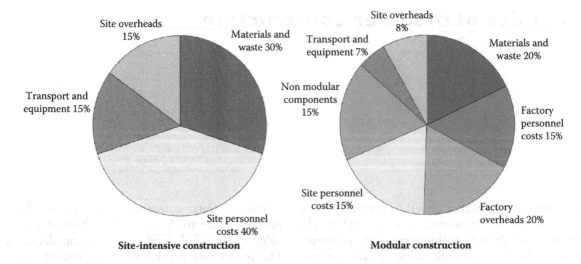

Figure 18.1 Comparison of breakdown of costs of site-intensive and modular construction of a multistorey residential building. (From National Audit Office, *Using Modern Methods of Construction to Build More Homes Quickly and Efficiently*, 2005.)

work is required in on-site finishing of these houses in the United States. He also states that for a factory output of 1000 modules per year, approximately 250 man-hours is required to produce one module, although the module size of 50 to 60 m² floor area is larger than in most European systems. This corresponds to an average level of productivity of 5 man-hours per m² floor area.

Background data on the economic benefits of modular and hybrid construction systems may be taken from a report by the National Audit Office: *Using Modern Methods of Construction to Build Homes More Quickly and Efficiently* (2005). Although, at the time of this report, modular systems were considered to be of similar as-built cost to more traditional on-site methods, savings of 7 to 8% due to the speed of construction are identified in it. It was also recognised that costs would reduce in larger projects or where modules are standardised in a range of similar projects.

18.2 ECONOMICS OF PRODUCTION

The economics of off-site manufacturing (OSM) technologies, and modular construction in particular, requires a significant production rate of relatively large-scale components whose materials, dimensions, and layout conform to an appropriate level of repetition. OSM also requires capital investment in the infrastructure of factory production, design development, product testing and certification, and overheads of a fixed factory facility.

The investment costs in factory production take into account the following fixed costs:

- Production equipment and infrastructure
- Factory running costs, including rental costs, heating, lighting
- Skilled personnel costs involved in manufacture

- Design and computer-aided design (CAD)/computer-aided manufacturing (CAM) facilities and training
- Storage and distribution facilities
- Downtime in manufacture

Modern automated factories for production of modular units may require an investment of the order of £5 to 10 million to set up, and these costs have to be amortised over a payback period of 5 to 10 years with a projected output of 1500 to 2000 units per year. The variety of modular solutions specified by clients and a changeable building market add to the complexity of achieving an economy of scale in manufacture. In comparison, the investment for a new automotive production line is often up to £500 million, but a successful car model has a typical annual production of over 50,000 and a 7- to 10-year production cycle.

Balanced against these fixed capital costs in more advanced manufacturing processes of the modules are the considerable improvements in productivity and savings in handling, use, and waste of materials in comparison to on-site construction. The improved quality of the manufactured units also saves in on-site checking and reworking.

Consider a typical modular production facility that requires an investment of £5 million and running costs of £1 million per year. Assume a 2-year build-up to full production and a 5-year production run of 1000 modules per year. It follows that the total investment cost per production unit would be approximately £2000. For a module of 30 m² floor area, this is equivalent to an investment of 10 to 15% of the ex-works cost of a typical unit. This is a significant investment, and must be balanced against other tangible savings, which are identified in the following section.

Even though the modular supplier will wish to use similar module designs and materials, the nature of the sector and the client specifications is that there will be significant differences between projects in terms of the modular design and manufacture. Therefore, the modular system should be capable of being flexible in its geometry and fitted-out components, but the basic framework and manufacturing process should be common to a wide range of projects.

An annual factory production of 1000–1500 modules may be broken down into 10 to 20 individual projects, which means that the design and management effort, and hence costs, is dependent on the number of projects rather than the number of modules. A typical modular hotel comprises 50 to 100 units, whereas a school may only comprise 6 to 20 units. Student residences are generally larger and can comprise 100 to 500 modules.

Consider a median project size of 100 modules, and so the design and setup costs associated with this project are added to the production cost of the modules. In this size of project, the design and management cost may amount to about 10% of the ex-works cost of a module. The modular supplier will produce the working drawings and manufacturing information for all the modules in a particular building, and this will often extend to cladding, roofing, and servicing. These costs should be balanced against the reduced fees of the architect and other consultants in a more traditional project (see later).

18.3 MATERIAL COSTS AND IMPROVED PRODUCTIVITY

The cost of materials is 30 to 35% of the ex-works cost of a module (or around 20% of the total building cost if the modular units are 60% of the total cost). This is less than the materials cost in an equivalent nonmodular building, despite the nature of the modular method of construction that requires a robust structural system for transportation and lifting. The savings in materials use mainly results from more precise ordering of materials to the sizes and quantities for a given project. Hence, board off-cuts and damage are greatly reduced. Savings in use of materials and wastage in modular construction could be as high as 15% of the total materials use in on-site construction. This greater efficiency in materials use is estimated as being equivalent to 3 to 4% savings in overall construction costs.

The productivity benefits in factory production lead to lower labour costs, although the manufacturer has to cover downtime in production, and the fixed personnel costs of directly employed production and design teams. It is difficult to determine the precise productivity benefits of performing similar construction techniques in the factory and on site. Also, the relative cost rate of the factory personnel and the equivalent site workers depends on the location of the factory and the site.

It may be estimated that for a median production of 100 modules on a given project, the productivity of working in factory conditions is increased by a factor of two for comparable tasks on the building site. For larger projects, productivity may increase by a factor of three. This will not translate directly into equivalent savings in labour cost, but productivity savings may be estimated at about 10% of the cost of the module as delivered from the factory in relation to the equivalent on-site construction.

18.4 PROPORTION OF ON-SITE WORK IN MODULAR CONSTRUCTION

Even in a highly modular project, a significant proportion of work is carried out on site in terms of installation of the modules and building of the nonmodular components. This can amount to 30 to 50% of the total cost of the project, and includes items such as enabling and ground works, concrete cores and stairs, cladding, roofing, central services installation, finishing and decorating, landscape work, and external infrastructure. The National Audit Report (2005) stated that the on-site work for a fully modular building could be typically broken down into foundations (5%), on-site services (8%), cladding and roofing (10%), and finishing work (7%), as a proportion of the total building cost.

Generally, modular companies try to minimise the number of on-site activities, but the ability to "parallel stream" design and manufacture modules with items such as enabling and ground works can be advantageous in terms of minimising construction programmes. The labour content in a given project may be considered to increase in proportion to the on-site activity.

Lifts and stairs may also be produced as modular components, although these are often constructed on site, unless the modular manufacturer has developed a close working relationship with specialist suppliers.

Data from the NAO Report (2005) on the relative performance of various methods of construction are summarised in Table 18.1. The NAO Report states that the other potential savings resulting from use of modular construction are as follows: For a brick-clad building, if bricklayers were on site for 44 days to construct a traditional building, they would be on site for only 20 days to construct the façade of a similar modular building. This is partly because the inner leaf of the wall is the module itself. Furthermore, scaffolding time would reduce from 11 weeks to 6 weeks, with commensurate savings in these costs. Furthermore, use of lightweight cladding attached to the modular units rather than brickwork will reduce these site requirements considerably.

The on-site labour activities are reduced dramatically for various levels of prefabrication, as shown in Table 18.2. According to the NAO data, the total

Table 18.1 Comparison of key time and cost factors in systems with various levels of prefabrication

Criteria	Traditional brick/ block construction	Panel (2D) construction	Hybrid panel and modular construction	Fully modular construction
Total construction period	100%	75%	70%	40%
Time to create weathertight envelope	100%	55%	50%	20%
On-site labour requirement (as a proportion)	100%	80%	70%	25%
Brickwork days on site	100%	45%	45%	45%
Proportion of total cost of on-site materials	65%	55%	45%	15%
Proportion of total cost of on-site labour	35%	25%	20%	10%
Proportion of total cost of off-site manufacture	0%	20%	35%	75%

Source:　National Audit Office, *Using Modern Methods of Construction to Build More Homes Quickly and Efficiently*, 2005.

Table 18.2 On-site labour effort in days (and savings in brackets) for typical terraced houses constructed by various methods

System	Foundations	Structure and walls	Finishes	Services	Total
Traditional brick- and blockwork	26	93	116	27	**262**
Panel construction (2D)	26	58 (35)	103 (13)	26 (1)	**207**
Hybrid panel and modular systems	26	40 (55)	108 (8)	23 (4)	**196**
Fully modular systems	26	29 (64)	13 (103)	0 (27)	**68**

Source:　National Audit Office, *Using Modern Methods of Construction to Build More Homes Quickly and Efficiently*, 2005.

Table 18.3 Summary of perceived risks for various forms of construction

Process stage	Risk description	Brick and block	Open panel	Hybrid	Modular
Planning	Unpredictable planning decisions			O	O
Preconstruction	Late appointment of supplier		O	●	●
Preconstruction	Lack of standardisation possible in the manufactured components		O	●	●
Detail design	Design changes after placement of order		O	●	●
Construction	Foundation inaccuracy affects installation		O	●	●
Construction	On-site components may be incompatible with manufactured components			O	●
Construction	Quality and accuracy problems	O			
Construction	Price fluctuations during construction	●			
Construction	Delays due to bad weather	●	O		
Construction	Lack of trade skills on site	●	O		
Construction	Service installation faults	●	O		
Construction	Health and safety hazards	●	O		
Occupation	Completed construction not to specification	●	O		
Occupation	Defects at handover or in liability period	●	O		

Source:　National Audit Office, *Using Modern Methods of Construction to Build More Homes Quickly and Efficiently*, 2005.

Note:　● = high risk, O = medium risk.

on-site labour effort in a fully modular building is only 26% of that of brick and block construction.

Risk management is a key feature of the use of off-site manufactured systems, which is summarised in Table 18.3, based on the NAO Report. Essentially, the key design decisions in modular construction have to be made at an earlier stage in the project, as late design changes are very difficult to incorporate when the modules are close to or in production. This implies a higher involvement of the modular supplier in the design process. This is identified by the NAO Report as a risk in terms of the client's ability to change the design at this stage, but this has to be balanced against the reduction in risks during the construction process, and the improved quality and in-service performance of modular systems.

18.5 TRANSPORT AND INSTALLATION COSTS

Transport costs in the UK may be taken as £500 to £800 per module for a 150-mile travel distance (each way to the site) provided a module is of a conventional size. However, additional costs will be incurred for an overnight stay in a holding area before delivery to the site, and for wider modules requiring additional notifications or police escort (see Chapter 16).

A 100- or 200-tonne capacity mobile crane would normally be required for installation on site, at a hire cost of up to £1000 per day. The crane size depends on the radius of lifting, as even a 200-tonne crane can only lift a small proportion of its capacity at its maximum extension. An average installation rate of 6 to 8 modules per day should be considered as realistic, although a rate of 10 to 12 may be achieved in the summer months.

The combined transportation and installation cost is therefore potentially up to £800 per module, which is about £25/m² for a module of 30 m² floor area. This is equivalent to about 4% of the ex-works cost of a module, or 2% of the overall construction cost of the building when expressed per unit floor area. Clearly, the larger the modules, the lower the relative transport costs. Two smaller modules may be transported on one lorry.

18.6 ECONOMICS OF SPEED OF CONSTRUCTION

18.6.1 Savings in site preliminaries

In site-intensive construction, site preliminaries may represent 12 to 15% of the total cost, and take into account:

- Management cost (related to the personnel required for management activities).
- Site huts and other facilities (in number and hire periods).
- Main contractor's equipment and craneage for materials handling and storage. (The mobile crane hired to install the modules is generally part of the modular package.)
- Construction time and programme (directly relates to personnel and hire costs noted above, as well as to cash flow).

Savings can be achieved due to the reduced number of site personnel (and hence costs and facilities needed), and the shorter construction programme (which is reduced by 30 to 50% in comparison to site-intensive construction). Based on estimates of site management costs and hire costs of site huts and equipment, the site preliminary costs for fully modular buildings may be taken as 7 to 8% of the total build cost, leading to a savings of 5 to 8% in comparison to site-intensive building projects.

18.6.2 Speed of installation

The benefits of speed of construction are inherent in modular systems of all types, and may be considered to be

- Reduced interest charges on borrowed capital
- Early start-up of the client's business, leading to earlier business or rental income
- Reduced disruption to the locality or existing businesses, mainly by the reduced build time, fewer deliveries, and site operations

These business-related benefits are clearly affected by the type of business, as the value of early completion may be different for projects that are time-constrained or where the disruption of the construction process can be quantified. A school or university building generally has to be ready for a particular time of the year, and so predictability in completion is a factor that often leads to the selection of modular construction. Hospitals also value reduced disruption and noise when using modular construction in the extension of existing facilities.

Snagging or improved quality control is reflected in cost savings of 1 to 2%, in addition to the risk reduction that is dependent on the client's business operation. Background testing of modular systems can also lead to efficiency gains by optimising performance and removing unnecessary waste in the design and manufacturing process.

18.6.3 Cash flow savings

Rogan (1998) carried out an assessment of light steel framing in housing and showed that the early return on the investment was due to reduced cash flow and capital employed. The study was later extended (2000) to a value and benefits assessment of modular construction. The tangible benefits of reduced interest charges due to a 6-month reduction in the overall construction programme can be 2 to 3% of the build cost.

It was calculated that for a medium-sized hotel, a savings of 1 month in the construction programme could amount to an income equivalent to 1% of the construction cost. Therefore, a 4-month reduction in a construction programme that is readily achievable using modular construction could be equivalent to a savings of 4% of the construction cost. This is often the determining factor in the decision to use modular construction in hotels.

18.7 SAVINGS IN DESIGN FEES

The overheads associated with the design of the modules and their interfaces with the rest of the building are borne by the modular supplier. Design includes structural design, service layouts, detailing, 3D computer modelling, production information, etc. Information produced by the modular supplier can be incorporated into the client's building information management (BIM) system and used to integrate with the other parts of the building.

Design and production costs will depend on the variability in the modules required for a particular project. Even a relatively simple project may comprise 4 to 12 different module configurations (including left- and right-handed modules). An allowance of the order of 10% of the ex-works cost of the modules may be made to meet in-house design and management costs in a typical project comprising 100 modules. This figure will increase significantly for smaller modular buildings unless the same, or similar, building type is repeated on other projects.

As a high proportion of the design work is carried out by the modular supplier, the cost of external consultants is reduced from typically 6 to 8% in traditional design and tender projects to 3 to 4%. However, the client's project architect is still responsible for overall coordination of the design of the building, and the client's structural and service consultants are responsible for the non-modular parts of the building.

18.8 SUMMARY OF POTENTIAL COST SAVINGS RELATIVE TO ON-SITE CONSTRUCTION

The relative costs of modular construction and traditional on-site construction should be divided into those related to the development cost and those that affect the construction cost. The client may be expected to save on consultant's fees and to gain from the financial savings of early completion, whereas the main contractor would gain from the reduced on-site costs and reduced risk by using off-site manufacture. These savings may be used to compare alternative constructional systems, and will vary depending on the scale of the project and the type of client's business.

The NAO Report (2005) estimated that the total financial savings when using modular construction due to these factors is typically 5.5% of the as-built costs. In the McGraw Hill Construction (2011) survey of prefabrication and modularisation in construction in the United States, it was reported that the average savings in projects with a high proportion of prefabrication was approximately 7.5%, although this included a range of prefabrication systems.

For the case of similar overall build costs, the savings in the use of modular systems may be summarised as follows:

Benefit of modular construction	Cost savings relative to site-intensive construction
Site preliminaries	5–8%
Client's consultant fees	3–4%
Snagging reduction	1–2%
Financial savings due to speed of construction	2–5%
Total savings as proportion of the total building cost	11–19%

REFERENCES

Buildoffsite. (2009). Your guide to specifying modular buildings: Maximising value and minimising risk. www.Buildoffsite.com.

Davis Langdon and Everest (now Aecom). (2004). Cost model of off-site manufacture. *Building*, 42.

McGraw Hill Construction. (2011). *Prefabrication and modularisation—Increasing productivity in construction.* SmartMarket report.

Mullen, M.A. (2011). *Factory design for modular home building.* Constructability Press, Winter Park, FL.

National Audit Office. (2005). *Using modern methods of construction to build more homes quickly and efficiently.*

Rogan, A.L. (1998). *Building design using cold formed steel sections—Value and benefit assessment of light steel framing in housing.* Steel Construction Institute P260.

Rogan, A.L, Lawson, R.M., and Bates-Brkjak, N. (2000). *Value and benefits assessment of modular construction.* Steel Construction Institute, Ascot, UK.

Chapter 19

Sustainability in modular construction

Sustainability, in the context of the planning of buildings, is quantified in terms of various measures of environmental, social, and economic performance. In the UK, offices and public buildings are assessed using BREEAM, and housing and residential buildings are assessed to the government's Code for Sustainable Homes (CfSH, 2010). Similar sustainability assessment procedures exist in other countries, such as Leadership in Energy and Environmental Design (LEED) in the United States.

The overarching requirements for environmental management systems are presented in BS EN IS 14001 (2004). PAS 2050 (2011) applies this methodology to environmental impacts and greenhouse gas emissions of manufactured products. It is compatible with other internationally recognised carbon footprinting methods.

This chapter reviews the features of off-site manufacture, and modular construction in particular, that contribute to improved sustainability in its widest sense. This also extends to embodied carbon and life cycle assessments (LCAs).

19.1 BENEFITS OF OFF-SITE MANUFACTURE ON SUSTAINABILITY

The off-site manufacturing (OSM) process in modular construction achieves many sustainability benefits that arise from the more efficient manufacturing and construction processes, the improved in-service performance of the completed building, and also the potential reuse at the end of the building's life. The sustainability benefits of OSM may be presented in terms of key performance indicators that are related to the construction process and in-service performance of modular buildings. These indicators are identified in Table 19.1 (based on Buildoffsite, 2014). Fully modular construction provides the highest level of off-site manufacture (as presented in Chapter 1), and therefore leads potentially to the highest sustainability benefits.

Modular units also have a significant residual value at the end of their design lives, and there is now considerable experience of modules being refitted out and reused elsewhere. The sustainability benefits in terms of in-service performance translate into potential scores under BREEAM or the Code for Sustainable Homes.

19.2 CODE FOR SUSTAINABLE HOMES (CFSH)

The Code for Sustainable Homes' assessment procedure is based on a number of environmental criteria, which are weighted separately and earn a percentage of available credits. The point scores are aggregated, but minimum scores must be satisfied in areas such as energy/CO_2, water savings, and material resources, in order to achieve an overall rating. Code level 3 is the default standard required for housing projects to energy standards of the 2010 Building Regulations (for the UK). Code level 6 is termed zero carbon and requires extensive use of on-site renewable energy technologies.

The available credits and weightings for each category of the assessment are presented in Table 19.2. The minimum reduction in the dwellings emission rate is linked to Part L of the Building Regulations (2010). The points required for each code category are shown in Table 19.3. However, a minimum standard is required in the energy and water categories in order to achieve each code rating.

The sustainability assessment of modular construction is presented in terms of various key performance indicators that link to the environmental criteria in the Code for Sustainable Homes. These criteria are as follows.

19.2.1 Energy and CO_2

The primary use of energy over the building's life is its operational energy that is due to heating (and in some cases cooling) and lighting. Low U-values of less than 0.2 W/m^2°C can be achieved in the external envelope of the modules by using "warm frame" construction, in which the majority of the insulation is placed externally to the unit. Modular buildings can also be designed and

Table 19.1 Sustainability benefits of off-site manufacture of modular systems in terms of construction process and in-service performance

Sustainability benefits of off-site manufacture as a construction process	Sustainability benefits of off-site manufacture in in-service performance
1. Social • Fewer accidents on site and in manufacture • More secure employment and training • Better working conditions in the factory • Reduced traffic movements to site • Less noise and disturbance during construction	1. Social • Acoustic insulation is improved due to sealed double-leaf construction • Improved finished quality and reliability • Future point of contact to the modular supplier • Modular buildings can be extended or adapted as demand changes
2. Environmental • Less pollution, including traffic, dust, noise, and volatile organic compounds (VOCs) • Less wastage of materials on site and in manufacture • More recycling of materials and use of materials with higher recycled content	2. Environmental • Improved energy performance by better airtightness and installation of insulation, hence, reduced CO_2 emissions • Renewable energy technologies can be built in and tested off site • Modular buildings can be "sealed" against gases, e.g., radon, and use on brownfield sites
3. Economic • Faster construction programme • Site preliminary costs are reduced • Less snagging and rework • Economy of scale in production reduces manufacturing cost • Higher productivity on site • Less site infrastructure and hire charges	3. Economic • Savings in energy bills, including by use of renewable energy systems • Longer life and freedom from in-service problems, e.g., cracking • Reduced maintenance costs • Modular buildings can be extended and adapted • Asset value of the modules can be maintained if they are reused

Table 19.2 Available credits to the Code for Sustainable Homes

Category	Credits	% of total
1. Energy and CO_2	29	36.4%
2. Water	6	9%
3. Materials	24	7.2%
4. Surface water runoff	4	2.2%
5. Waste	7	6.4%
6. Pollution	4	2.8%
7. Health and well-being	12	14%
8. Management	9	10%
9. Ecology	6	12%

Table 19.3 Code for sustainable homes ratings for various code * levels

Code level		Total score	Minimum reduction DER
1	(*)	36	10%
2	(**)	48	18%
3	(***)	57	25%
4	(****)	68	44%
5	(*****)	84	100%
6	(******)	90	Zero carbon

Note: DER = dwelling emission rate.

manufactured to be very airtight by the use of additional membranes and sheathing boards.

The code does not address the embodied energy in its materials over the design life span, which can be important as the operational energy of the building reduces. The embodied carbon is the equivalent amount of CO_2 produced in the manufacture of the materials, which is discussed in Section 17.6.

19.2.2 Materials

Modular construction is efficient in the use of materials by efficient ordering to the sizes and quantities required, and results in less waste. The Building Research Establishment (BRE) *Green Guide to Specification* (2009) measures the environmental impact of building systems according to various criteria, including embodied carbon, waste, recycled content, etc. The ratings are presented on a scale of A* (highest) to E (lowest). The lightweight building elements in modular construction conform to the A*, A, or B ratings.

In terms of materials use, 98% of all steel is recycled after its primary use, and 50% of current steel manufacture in Europe comes from recycled steel (scrap). No structural steel is sent to landfills. For concrete modules, the more repetitive factory approach ensures efficient placement and reuse of concrete formwork, when compared with in situ construction. Reinforcement used in concrete modules is almost 100% manufactured from recycled steel.

Savings in foundation sizes can be significant when lightweight modular units are used, which is very important when building on brownfield sites and poor ground.

Furthermore, modules are reusable and their asset value is maintained. Potentially, the modules can be refurbished and their life is extended considerably (see Chapter 8).

19.2.3 Waste

Waste in on-site construction arises from various sources:

- Over-ordering to allow for off-cuts
- Damage and breakage, and losses on site
- Rework due to errors on site

According to the Building Research Establishment (BRE), the construction industry average for material wastage on site is 10%, although this varies with material (Smartwaste). In comparison, for modular construction, waste is minimised in the manufacture and installation process. All off-cuts are fully recycled in the factory.

WRAP has carried out various case studies on waste reduction and waste recycling, including a study (WRAP, 2008) of the use of modular construction in the regeneration of the Woolwich barracks in London.

In concrete modules, wastage in concrete is minimised in the batching and placing of concrete, part of a single operation, compared with in situ concrete, which is ordered in from a ready-mix company potentially some distance away from the site. It is commonplace to over order in situ concrete by at least 10% to ensure adequate supply for a given on-site pour. A Hong Kong study by Jaillon and Poon (2008) showed that, on average, a production of precast concrete panels and units leads to a reduction of 65% in construction waste in comparison to in situ concrete.

Packaging of separate components is also minimised in modular construction, and the weather-protective shrouds around the modules are left in place and assist in providing long-term durability.

19.2.4 Water

Construction of light steel and timber modular systems is essentially a "dry" process in the factory and on site.

In concrete construction, water is used in the manufacture of concrete modules, but less than would be used if the concrete was placed in situ due to the more controlled factory environment.

19.2.5 Pollution

Much less noise, dust, and noxious gases are generated on site when using modular construction systems of all types. In highly prefabricated construction systems, transportation of materials to the site is reduced by around 70% in comparison to brick- and blockwork construction, which leads to a consequent reduction in deliveries to the site and local traffic pollution.

Raw materials are delivered in bulk to the module factory, to the correct quantities and sizes, which is more efficient than the multiple smaller deliveries to site.

19.2.6 Management

Site management is much improved by "just in time" delivery of the modules and minimal storage of materials on site. Installation teams are highly skilled, efficient, and productive. Noise and other sources of disturbance are also minimised, which is important in terms of considerate construction.

As noted above, site deliveries and traffic due to construction activities are also reduced relative to more traditional ways of building, based on data in a recent National Audit Office report (see Section 18.4).

Also, an increasingly important part of building information management (BIM) systems is that the *electronic* design model of the modular structure and layout is available to all members of the design and construction team and can be retained by the client for future records.

19.2.7 Performance improvements

Modular units are strong and robust to damage. Steel and concrete are non-combustible and do not add to the fire load. High levels of acoustic isolation and thermal performance can be achieved, and precast concrete is also inherent in terms of its thermal mass and security.

Shrinkage or long-term movement on site is reduced by building in dry factory conditions, and "callbacks" to rectify errors and snagging are largely eliminated by the checking of the modules before delivery to the site. The durability of galvanised steel sections has been assessed in site trials, and a design life of over 100 years is achieved in warm frame applications, where the majority of the insulation is placed outside the light steel frame (refer to SCI P262; Lawson et al., 2009).

19.2.8 Adaptability and end of life

The open-plan space of the modules can be fitted out and serviced to suit the user's requirements, and modular buildings can be later disassembled and reused.

Concrete modules are extremely robust, and it is possible to make attachments by chemical anchors or expanding bolts. Open-ended concrete modules are more flexible and adaptable. The modules are designed for the easy access and replacement of non-structural components, such as electrics, plumbing, and furnishings.

A recent example of the reuse of modules was in a training centre at Freeman Hospital in Newcastle-upon-Tyne. The scheme comprised 10 preowned steel-framed modules that were recycled and refurbished for the project, enabling it to be delivered in just 11 weeks from receipt of order to handover. It was also claimed that using refurbished modules generated less than 10% of the embodied carbon and used less than 3%

of the energy during construction, compared to a site-constructed building of equivalent size.

19.2.9 Social responsibility

All steel components are stamped with their project title, date, and component number, and are traceable to their original source. The galvanised strip steel or hot-rolled steel sections that are used have environmental product declarations, and have guaranteed high-strength properties and suffer no long-term deterioration.

Module manufacturing and installation processes require a high level of skills and training with excellent job opportunities. From the Health and Safety Executive data, factory-based processes, including modular manufacture, are five times safer than construction processes in terms of the number of reportable accidents. Modular manufacture contributes to a clean and safe working environment with a good employment and training history.

Risks associated with working at height are also avoided, thus improving workers' safety on site. In one study, the safety risk associated with manufacture and installation of precast concrete was up to 63% lower than with conventional on-site concrete construction (Jaillon and Poon, 2008).

On-site noise and disruption is minimised, and so modular construction does not affect the neighbourhood adversely during the construction process compared with more traditional construction processes.

19.3 BACKGROUND STUDIES ON SUSTAINABILITY

Various background studies have been carried out to evaluate the environmental and sustainability benefits of OSM. In the National Audit Office (NAO) report *Using Modern Methods of Construction to Build More Homes Quickly and Efficiently*, the types of construction systems investigated were 2D panels, hybrid 2D and 3D, and fully 3D construction, and they were compared to more traditional construction.

A key indicator for the completion of a building was the time to create a weathertight envelope, which, for fully modular construction, reduces to 20% of the time required for traditional brick and block construction. On-site labour time (and hence costs) is reduced to 25 to 75% of traditional construction, depending on the level of prefabrication.

The BRE SmartLife project (Cartwright et al., 2008) compared the site productivity and materials waste of brick- and blockwork construction, light steel framing, timber framing, and insulated concrete formwork for three similar housing developments in Cambridge. It

was shown that the light steel system was the fastest to construct, had the highest productivity, and generated the least waste while being comparable in cost to traditional construction. It may be expected that the benefits of modular construction would be even greater.

19.3.1 Sustainability assessment of a high-rise modular building

In 2009–2010, three residential buildings in Wolverhampton (see case studies) of 8, 9, and 25 stories height were monitored to ascertain the rate of installation of the modules, the site productivity (defined by the number of workers per unit floor area completed), the number of major deliveries of building components and materials, and the waste created. The 8-storey building is shown during construction in Figure 19.1. The overall construction period for this project of 25,000 m^2 floor area was only 15 months (from foundation level), which represented an estimated reduction of 12 months relative to an equivalent in situ concrete project.

The rate of installation ranged from 28 to 49 modules per week (at an average of 7.5 per day) during the winter months. The average number of site personnel, including management, was 52 over the yearlong construction period (of which 5 were involved in site management), and so the rate of completion of the building was approximately 18 m^2 floor area per person per week. It was estimated by the main contractor that the personnel required for an equivalent in situ concrete construction with blockwork walls would be over 100, and so the reduction in site personnel was about 50%. The site personnel were mainly required for the construction of the concrete podium and cores, and installation of the cladding.

The average number of major deliveries to the site per day (excluding the modules) was 6 (or 14 including the delivery of the modules), which represents a reduction of over 60% relative to the deliveries of the high volume of concrete and blockwork required for in situ construction. The number of waste skips sent for disposal averaged two per week, which was an estimated reduction of over 95% relative to in situ construction. This led to considerable savings in disposal costs for the main contractor.

The sustainability benefits achieved on this large residential project are summarised by Lawson et al. (2012), as follows:

Summary of performance achieved

- 7.5 modules installed per day, equivalent to 190 m^2 floor area
- Reduction of 12 months in construction period (40% overall reduction in construction period)

Figure 19.1 Twenty-five-storey modular residential building in Wolverhampton used in the study of sustainability.

- Site personnel reduced by 50% (on average, 52 workers were employed continuously throughout the 25,000 m² project, including 5 site management staff)
- Site waste sent to disposal reduced by over 95%
- Deliveries to site reduced by 60%

19.3.2 Sustainability assessment of Open House modular system in Sweden

A sustainability assessment has been made by the Swedish Steel Construction Institute of a residential light steel modular system called Open House, which has been used in housing developments in Denmark and southern Sweden (Birgersson, 2004; Lessing, 2004). The building project studied was the construction of 1200 apartments in Annestad near Malmö, Sweden, shown in Figure 19.2. The study (Widman, 2004) included an assessment of the following sustainability issues:

- Materials use and resources
- Operational and embodied energy
- Waste and recycling in manufacture and construction
- Construction efficiency and safety
- Flexibility and adaptability in use
- Social benefits arising from the off-site manufacturing process

The Open House system is presented in Chapter 10 and in the case studies. The materials used were compared to a national average modern residential building in Sweden, which is a concrete structure with blockwork infill walls. The main structural components in a modular apartment weighed 148 kg/m² floor area, excluding the façade, of which steel usage was 41 kg/m² (or 28% of the total). A typical residential building in Sweden has a material weight of about 982 kg/m², excluding the façade, which is over six times higher than in the modular system. It was found that the factory production of a fully equipped Open House module took about 65 man-days, or approximately 15 man-hours per m² floor area.

This study also showed that site wastage was very small when using modular construction, and most waste was recycled in factory production. Also, the recycled content of the Open House system was 67 kg/m², which corresponded to 45% of the total material use. The recycled content of the comparative national average building is about 49 kg/m², which corresponds to 5% of the weight of the materials used (façade excluded).

The national average operational energy use for modern multistorey residential buildings in Sweden is 160 kWh/m², which includes space heating, hot water, and lighting. In-service measurements of the Open House system indicated that the actual energy use was 120 kWh/m². Over 50 years, the operational energy use is therefore calculated as 6 MWh/m² for the Open

Figure 19.2 Completed Open House project in Malmo, Sweden.

House system, which is 25% less than for the reference building.

The production of the main components in a national average residential building structure leads to an embodied energy of 333 kWh/m² floor area, including the façade, or 245 kWh/m² when calculated without the façade. Over a 50-year life, this is equivalent to about 3% of the total operational energy that is consumed in heating, lighting, etc. The calculated embodied energy figures for the Open House system were 215 kWh/m² including the façade, and 178 kWh/m² without the façade. Therefore, the embodied energy using the modular system was calculated to be 28% lower than the national average reference building. As part of the embodied energy calculation, the energy used in transport of the building materials and off-site components was calculated as 6 kWh/m² (façade excluded) for the modular building, which was only one-third of that of the reference building.

19.4 EMBODIED ENERGY AND EMBODIED CARBON CALCULATIONS

Embodied energy and embodied carbon calculations for completed buildings are increasingly requested by clients. The key references for embodied energy and embodied carbon are those by Hammond and Jones of the University of Bath (2008a) and by BSRIA (2011).

Data for common materials used in buildings are given in Table 19.4. The embodied carbon parameter is defined as the weight of the equivalent CO_2 emissions resulting from the extraction, refining, and manufacture of the building components in their respective materials and does not include the CO_2 emissions in the construction process. Credits are taken for recycled content and for end-of-life recycling, although end-of-life credits are much less given the long cycle in building use and refurbishment.

The approximate breakdowns of embodied energies and embodied carbon for a typical detached house are presented in Table 19.5 (summarised from Hammond and Jones (2008b)). The concrete ground slab and brickwork cladding amount to 37% of the total embodied energy and to 58% of the embodied carbon. Hammond and Jones quote a median embodied energy of 5200 MJ/m² and 360 kg CO_2/m² floor area for a single-family house, increasing to 480 kg CO_2/m² floor area for a 4-storey residential building. Sansom and Pope (2012) quote a comparative figure of 450 to 480 kg CO_2/m² floor area for an office building.

19.4.1 Embodied carbon in transportation

The embodied carbon in various forms of road transport is presented in Table 19.6, depending on the types and size of HGV lorry. These data published by Department for Environment, Farming, and Rural Affairs (DEFRA)

Table 19.4 Embodied energies and embodied carbon of building materials

Material	Typical density (kg/m³)	Embodied energy (MJ/kg)	Embodied carbon (kgCO₂/kg)
Bricks	1800	3	0.24
Cement mortar	2200	1.3	0.22
Concrete—structural	2400	1.0	0.15
Concrete (with fly ash)	2400	0.9	0.13
Reinforced concrete (1% rft)	2450	1.9	0.22
Precast concrete	2400	1.5	0.18
Screed (sand and cement)	1200	0.6	0.07
Autoclaved aerated concrete (AAC) blocks	600	3.5	0.28–0.37
Concrete blockwork (medium)	1400	0.59	0.063
Concrete blockwork (heavy)	2000	0.72	0.088
Clay tiles	1600	6.5	0.48
Steel (sections)	7850	21.5	1.53
Galvanised steel	7850	22.6	1.54
Stainless steel	8000	56.7	6.15
Reinforcement	7850	17.4	1.4
Plywood	800	15	0.81
Timber joists/frame	700	10	0.46
Plasterboard	850	6.75	0.39
Oriented strand board (OSB)	850	15	0.6
Mineral wool	30–140	16.6	1.28
Expanded polystyrene (EPS) insulation	25	88.6	3.29
Polyurethane (PUR) insulation board	30	101	4.26
Glass fibre	45	28	1.54
Glass (4 mm)	2500	15	0.91

Source: Hammond, G. P., and Jones, C. I., *Inventory of Carbon and Energy (ICE)*, Version 1.6a, University of Bath, Bath, UK, 2008.

Table 19.5 Breakdown of embodied energies and carbon in a typical house

	Embodied energy % of total	Embodied CO₂ % of total
Concrete and bricks	37%	58%
Insulation	12%	7%
Clay tiles	7%	5%
Steel elements	6%	5%
Wood elements	6%	3%
Windows and glass	5%	3%
Plastics	3%	2%
Plasterboard and plaster	2%	2%
Copper pipes and wires	2%	2%
Other	20%	16%

Source: Hammond, G. P., and Jones, C. I., *Proceedings of the Institution of Civil Engineers: Energy*, 161(2), 87–98, 2008.

Table 19.6 CO₂ emission factors per tonne kilometers travelled for HGV road freight

Body type	Gross vehicle weight	kgCO₂ per tonne km
Rigid	>3.5–7.5 t	0.59
Rigid	>7.5–17 t	0.33
Rigid	>17 t	0.19
All rigid	**UK average**	**0.27**
Articulated	>3.5–33 t	0.16
Articulated	>33 t	0.082
All articulated	**UK average**	**0.086**
All HGVs	**UK average**	**0.132**

19.4.2 Embodied carbon study

Using the above data, a study is made of the materials use and embodied carbon in a typical light steel module and a precast concrete module with an open base. The modules are 7.2 m long by 3.6 m wide (26 m² floor area) and are designed for use in a 4-storey residential building. Common aspects, such as the cladding, windows, insulation, finishes, services, internal partitions, etc., are not included, and so the study concentrates on the structural fabric of the modules. Mineral wool insulation between the C sections and plasterboard are

(2008) in the UK are expressed in kgCO₂ per km travelled for every tonne weight of the lorry and its load. When using modular construction, it is estimated that the number of deliveries to the site are reduced by over 70% when compared to site-intensive construction. This is combined with fewer daily journeys by site workers, etc., which are not normally calculated in the embodied carbon assessments.

included for the light steel module in its finished state. The concrete module is assumed to be fair-faced concrete with a skim plaster coat. In both cases, additional external insulation is required to meet thermal requirements, but this is not included in the assessment.

The data and results for a light steel module are presented in Table 19.7. The total quantity of steel used in the module is about 1 tonne, which is equivalent to 38 kg/m² floor area. The self-weight of the fabric of the module without its finishes and services is approximately 150 kg/m², and the estimated finished weight is around 300 kg/m². Therefore, a completed module weighs about 7 tonnes. The embodied carbon, including transport, is estimated as 129 kg/m² floor area, which is equivalent to 84% of the weight of the structural fabric of the module.

The data and results for a precast concrete module are presented in Table 19.8. The self-weight of the module without finishes and services is 935 kg/m², and the estimated finished weight is around 1000 kg/m², which is four times that of the light steel module. Therefore, the completed module weighs about 27 tonnes. The embodied carbon is 184 kg/m², which is 42% more than the light steel module, but is equivalent to only 19% of the weight of the concrete module.

The embodied energy in the materials of the light steel module was, however, 70% higher than that of the concrete module. This is due to the higher energy required to produce the light steel components. In modern methodologies, the embodied carbon is a better measure because it takes account of the carbon intensity of the energy sources that are used in manufacture of the materials.

The embodied carbon used in transport assumes a 200 km journey of the module from the factory to the site, and the same return journey for the articulated lorry of 3.5 tonnes of unladen weight. This shows that the embodied carbon in transport is 12% of the

Table 19.7 Total quantities in materials and embodied carbon in the fabric of a light steel modular unit

Component	Total weight of materials (kg)	Embodied energy MJ/kg	Embodied carbon kgCO₂/kg	Total embodied energy MJ	Total embodied carbon kgCO₂
Walls and ceiling: 100 × 1.6 Cs at 600 mm centres	680	22.6	1.54	15,370	1047
Floor: 200 × 1.6 Cs at 400 mm centres	320	22.6	1.54	7230	493
OSB sheathing boards	860	15	0.62	12,900	533
Plasterboards (2 layers of 15 mm boards on walls and ceiling)	1850	6.79	0.39	12,560	721
Mineral wool between the C sections	250	16.6	1.28	4150	320
Transport to site (200 km journey)	6500 kg including finishes	—	0.086 kgCO₂/ tonne km	Not included	232
Totals	3960 kg			52.2 GJ	3346 kgCO₂
Total per m² floor area	153 kg/m²		or	2014 MJ/m²	129 kgCO₂/m²

Note: Data for 3.6 m wide by 7.2 m long module.

Table 19.8 Total quantities in materials and embodied carbon in the fabric of a precast concrete modular unit

Component	Total weight (kg)	Embodied energy MJ/kg	Embodied carbon kgCO₂/kg	Total embodied energy MJ	Total embodied carbon kgCO₂
Walls: 125 mm thick concrete panels	15,400	1.0	0.15	15,400	2310
Ceiling: 150 mm thick concrete slab	8400	1.0	0.15	8400	1260
T10 reinforcing bars at 200 mm centres on both faces of all walls	315	17.4	1.4	5480	441
T12 reinforcing bars at 200 mm centres on both faces of ceiling	125	17.4	1.4	2180	175
Transport to site (200 km journey)	26,800 kg including finishes	—	0.086 kgCO₂/ tonne km	Not included	581
Totals	24,240 kg			31.5 GJ	4767 kgCO₂
Total per m² floor area	935 kg/m²		or	1214 MJ/m²	184 kgCO₂/m²

Note: Data for 3.6 m wide by 7.2 m long module.

total for a concrete module and 7% for a light steel module, and so transport is not a negligible factor in these assessments.

Factors that are not included in this assessment are the wastage of materials and the construction operations. The embodied carbon of the on-site construction process is generally taken as about 5% of that of the materials used, and this may be considered to reduce to 2 to 3% in modular construction because of its lower use of machinery and equipment, site huts, and daily travelling to and from the site. The energy required to lift and install the modules is relatively small.

It may be concluded from this study that the embodied carbon in the structural fabric of a light steel module is 30% less than that in a concrete module, when expressed per unit floor area. Both methods of construction possess lower embodied carbon than the equivalent in situ construction, partly because of the reduced wastage in materials. A more sophisticated analysis may take account of the variable elements in the external insulation, cladding, and foundations, and their wastage rates, and also the components that are common to both forms of construction.

CASE STUDY 41: TWENTY-FIVE-STOREY STUDENT RESIDENCE, WOLVERHAMPTON

Twenty-five-storey block A building under construction.

Eight-storey block near completion.

Victoria Hall Ltd. procured three multistorey student residences for Wolverhampton University, which were built in modular construction. At 25 storeys, one block is currently the tallest modular building in the world. The modules were manufactured by Vision, and the contractor was Fleming. A total of 820 modules were installed in 9 months, and the construction period was only 15 months to handover in August 2009.

The project is located next to the main railway line north of the centre of Wolverhampton. Importantly, the reduction in on-site activities and storage of materials by using modular construction was crucial to the planning of this city centre project. Block A is 25 storeys high on its southern side and 18 storeys high on its northern side. The modules are ground supported on a reinforced concrete slab. The 3rd, 7th, 12th, and 18th floors are set back on one side and form a cantilever to the floors above. This cantilever is supported by a steel frame.

A feature of the modular construction was the use of integral corridors manufactured as part of the modules, which created a weathertight envelope for the group of modules. All horizontal loads were transferred in-plane by the modules, and overall stability was provided by concrete cores. All vertical loads are resisted by the walls of the modules.

The modules have a concrete floor cast within 150 mm deep parallel flange channel (PFC) sections, and 60 × 60 square hollow section (SHS) posts at 600 mm centres form the load-bearing side walls. Larger rectangular hollow sections (RHSs) are used on the lower modules in the high-rise block. The modules in this project were delivered up to 4 m wide and 8 m long and included a central 1.1 m wide corridor. The study bedrooms were typically 2.5 m wide by 6.7 m long, but the communal rooms were up to 4.2 m wide.

Installation of the modules started in block C in October 2008 and was completed in only 6 weeks. Installation of modules in block B took a further 6 weeks. For block A, a tower crane with a 30-tonne capacity at 20 m radius was used to install the modules, and was attached to the modules for stability. Installation of modules in this block was completed in 17 weeks.

Horizontal ties were provided at the corners of the modules and at intermediate points along their sides. The modules transferred the required wind forces to cast in steel plates within the reinforced concrete core. Cladding was ground-supported brickwork at the lower levels and a mixture of insulated render, composite panels, and rain screen metallic panels at the upper levels. The lightweight cladding was installed using hoist towers attached to the modules.

REFERENCES

Building Regulations England and Wales. (2010). *Conservation of fuel and power*. Approved Document Part L. www.planningportal.gov.uk/buildingregulations/approveddocuments.

Building Research Establishment. (2009). *Green guide to specification*. www.bre.co.uk/greenguide.

Building Research Establishment Environmental Assessment Method (BREEAM). www.breeam.org.

Birgersson, B. (2004). *The Open House 3D modulus system*. SBI Report 229:3. Stålbyggnadsinstitutet, Stockholm.

British Standards Institution. (2004). *Environmental management*. BS EN ISO 14001.

British Standards Institution. (2011). *Specification for the assessment of the life cycle greenhouse gas emissions of goods and services*. PAS 2050.

Building Services Research and Information Association. (2011). *Embodied carbon—The inventory of carbon and energy (ICE)*.

Buildoffsite. (2013). Offsite construction: Sustainability characteristics. London, UK. http://www.buildoffsite.com.

Cartwright, P., Moulinier, E., Saran, T., Novakovic, O., and Fletcher, K. (2008). *Smart life—Lessons learned: Building research establishment*. BR 500.

Code for Sustainable Homes. (2010). Technical guidance. www.gov.uk/government/publications.

Department for Environment, Farming and Rural Affairs (DEFRA). (2008). *Guidelines to Defra's GHC conversion factors—Methodology for transport emission factors*. www.defra.gov.uk.

Hammond, G.P., and Jones, C.I. (2008a). *Inventory of carbon and energy (ICE)*. Version 1.6a. University of Bath, UK.

Hammond, G.P., and Jones, C.I. (2008b). Embodied energy and carbon in construction materials. *Proceedings of the Institution of Civil Engineers: Energy*, 161(2), 87–98.

Jaillon, L., and Poon, C.S. (2008). Sustainable construction aspects of using prefabrication in dense urban environment: A Hong Kong case study. *Construction Management and Economics*, 26(9), 953–966.

Lawson, R.M. (2007). Sustainability of steel in housing and residential buildings. The Steel Construction Institute, P370.

Lawson, R.M., Ogden, R.G., and Bergin, R. (2012). Application of modular construction in high-rise buildings. American Society of Civil Engineers. *Journal of Architectural Engineering*, 18(2), 148–154.

Lawson, R.M., Way, A., and Popo-ola, S.O. (2009). *Durability of light steel framing in residential buildings*. Steel Construction Institute P262.

Leadership in Energy and Environmental Design (LEED). www.usgbc.org.

Lessing, J. (2004). *Industrial production of apartments with steel frame. A study of the Open House System*. SBI Report 229:4. Stålbyggnadsinstitutet, Stockholm.

National Audit Office. (2005). *Using modern methods of construction to build more homes quickly and efficiently*.

Sansom, M.R., and Pope, R.J. (2012). A comparative embodied carbon assessment of commercial buildings. *The Structural Engineer*, October 2012, pp. 38–49.

Smartwaste. www.smartwaste.co.uk.

Widman, J. (2004). *Sustainability of modulus construction. Environmental study of the Open House steel system*. SBI Report 229:2. Stålbyggnadsinstitutet, Stockholm.

WRAP. (2008). Woolwich single living accommodation modernisation (SLAM) regeneration. www.wrap.org.uk.

Index